CYCLONE WARRIORS

The Australian Defence Force and Cyclone Tracy
December 1974 – June 1975

Dr Tom Lewis

Foreword by Sir Peter Cosgrove

Cyclone Warriors
The Australian Defence Force and Cyclone Tracy
December 1974 – June 1975

Dr Tom Lewis

ISBN 9780975642313

First published 2024 by Avonmore Books

Avonmore Books
PO Box 217
Kent Town
South Australia 5071
Australia

Phone: (61 8) 8431 9780

avonmorebooks.com.au

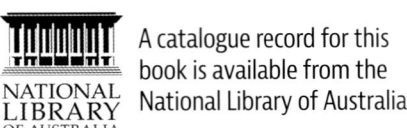

A catalogue record for this book is available from the National Library of Australia

Cover design & layout by Diane Bricknell

CONTENTS

ABBREVIATIONS 4

ACKNOWLEDGEMENTS 5

FOREWORD 6

BY THE SAME AUTHOR 8

EXPLANATORY NOTES 9

MAPS 11

INTRODUCTION – CYCLONE TRACY AND THE ARMED FORCES 13

CHAPTER 1 – THE ARMED FORCES IN DARWIN PRIOR TO TRACY 14

CHAPTER 2 – THE CYCLONE STRIKES – THE AUSTRALIAN DEFENCE FORCE STORY 17

CHAPTER 3 – THE CONCEPT OF RESCUE, AND MILITARY COMMAND 31

CHAPTER 4 – THE AIR FORCE STORY 35

CHAPTER 5 – THE NAVY STORY 47

CHAPTER 6 – THE ARMY STORY 66

CHAPTER 7 – UNIFORMS APART FROM THE ADF 83

CHAPTER 8 – STRETTON DEPARTS 91

CHAPTER 9 – WHY NO MEDAL FOR CYCLONE TRACY? 94

CHAPTER 10 – TRACY CONTROVERSIES 97

CONCLUSION – CYCLONE TRACY AND THE ARMED FORCES 115

APPENDIX – THE CURIOUS CASE OF MAJOR GENERAL STRETTON VERSUS THE ARMY 115

LIST OF WORKS CONSULTED 121

INDEX OF NAMES 127

ABBREVIATIONS

ABC	Australian Broadcasting Corporation
ADF	Australian Defence Force
ARA	Australian Regular Army
ATTU	Air Transportable Telecommunications Unit
CO	Commanding Officer
CQMS	Company Quarter Master Sergeant
CSM	Company Sergeant Major
DCNS	Deputy Chief of Naval Staff
DGNDO	Director General Natural Disasters Organisation
DMR	Department of Main Roads
FOCAF	Flag Officer Commanding the Australian Fleet
HMAS	His/Her Majesty's Australian Ship
HMS	His/Her Majesty's Ship
LCH	Landing Craft Heavy
MACC	Military Aid to the Civil Community
NDO	Natural Disasters Organisation
NOCNA	Naval Officer Commanding Northern Australia
NT	Northern Territory
OAM	Order of Australia Medal
OC	Officer Commanding
RAAF	Royal Australian Air Force
RAF	Royal Air Force
RAN	Royal Australian Navy
RANR	Royal Australian Navy Reserve
RAR	Royal Australian Regiment
RNZAF	Royal New Zealand Air Force
USAAF	United States Army Air Force
WWII	World War Two
WRAAF	Women's Royal Australian Air Force
WRANS	Women's Royal Australian Naval Service

ACKNOWLEDGEMENTS

In alphabetical order:

- Ms Kaylene Anderson, as always
- Mr Jared Archibald OAM, History Curator, Museum and Art Gallery of the NT
- Mr Michael Claringbould, Military Aviation Historian
- General the Honourable Sir Peter Cosgrove AK AC (Mil) CVO MC (Retd)
- Mr Col Coyne, President 37SQN (RAAF) Association
- Darwin City Council
  - Ms Bernadett Howison
  - Mr Thomas Schelling
- Nigel Daw, South Australian Aviation Museum
- Ms Linda Fazldeen AM
- Group Captain David Fredericks, Director History and Heritage Services, Royal Australian Air Force
- Ms Katherine Hamilton
- Mr Earl James AM
- Ms Cathy Johnston
- The Honourable Daryl Manzie AM and Mrs Maureen Manzie
- Mr Paul McAlonan of the Army History Unit
- Northern Territory Police, Fire and Emergency Services
  - Ms Genevieve Reid, Information Officer
- Dr Peter Pedersen AM
- Mr Ian and Ms Jennifer Richards, with thanks for the title idea
- Royal Australian Navy Seapower Centre
  - Commander Alastair Cooper RAN
  - Mr Petar Djokovic, Historian, Naval History Section
- The Honourable Mr Peter Styles
- The Honourable Mr Grant Tambling AM and Mrs Sandy Tambling
- Dr Peter Williams, Military Historian
- Ms Sharon Yang, Royal Australian Air Force
- Mr Charlie Ward

To Chris Allison and Christine Barden, both Territorians

FOREWORD

I was fascinated and grateful to be asked by Tom Lewis to read and contemplate writing the foreword for *Cyclone Warriors*. In 2024, December 25 will mark fifty years since Cyclone Tracy, having smashed places like Melville Island, came ashore over Darwin city, that remarkable gateway to Australia's north.

I was first, fascinated to read Tom's insightful accounts of how Australian Defence Force people fared through the hugely dangerous cyclone and how they reacted in the aftermath; how in Darwin and all-around Australia, the men and women of the ADF enthusiastically responded to the crisis, in many cases anticipating urgent government direction, getting on to do needful things. Military forces of the three services have been engaged in similar support to the wider community throughout Australia's modern history but when you read Tom's account, you'll join me in the persuasion that Cyclone Tracy built an enduring platform of expectation both within the ADF and in the wider national community for the way the public hopes the ADF will act in such situations.

I mentioned "gratitude" above. I was one of the thousands of uniformed men and women (both those serving on postings in Darwin and from "down south") working on the rehabilitation of Darwin in the near aftermath of the cyclone. I vividly remember flying into Darwin, part of a two-person reconnaissance team, to set up arrangements for the first tranche of the infantry-battalion group (mostly infantry, plus a squadron of Armoured Corps soldiers and some other corps specialists). We were essentially a labour force. As we flew into Darwin on an RAAF C-130 transport aircraft, looking out the windows there were two indelible, unforgettable sights. First, the city of Darwin reminded me of photos I'd seen of the A-Bomb struck cities of Hiroshima and Nagasaki. Several thousand homes and small businesses which were high set, had been wiped away leaving the wreckage strewn downwind. Secondly, extraordinarily uplifting, was the sight in the harbour of a goodly part of the Navy's fleet, with ships' boats shuttling backwards and forwards to bring hope and help to those enormously needy people overwhelmed by the wreckage. On landing at RAAF Darwin, the continued operation to take needy people to places of shelter and support in the south and bring necessary people and stores in was being done with great energy and skill by our Air Force men and women.

My soldiers and I were there for about seven weeks. The cleanup work was backbreaking, never-ending, vastly uncomfortable but every soldier performed magnificently, knowing it was needed, and what Australia wanted and expected. That shines through in Tom's account. That's why I am grateful and so will they be.

As a postscript to this personal reflection, I vividly recall on my first arrival into Darwin city, reporting to Captain Eric Johnston, who was in charge of the naval cleanup teams ashore, at the grandly named Admiralty House. His imposing physique and huge energy ran through everybody in his vicinity. He had officers and sailors charging in all directions. I have had many experiences throughout my career with the men and women of our Navy – that few days of the Navy/Army overlap made me unforgettably proud.

I think readers will also be gratified that Tom has included insights into the Herculean work of the NT police in their community facing duties over the whole period of the cyclone and its aftermath.

Tom does not shrink from discussing some of the classic controversies arising from the events of December 1974, but I commend him for bringing them to us away from the central account. It's in the nature of humanitarian crises that there will be rumour and legend but at its core, there was enough provable, observable tragedy to set Cyclone Tracy in the front rank of natural disasters in our lifetime.

In closing this foreword, I congratulate Avonmore Books for their role in helping Tom bring this story to the Australian people and beyond.

Of course, primarily my congratulations and admiration go to Doctor Tom Lewis, OAM, for his absorbing, forensic and valuable insights into the way the Australian Defence Force rallied to meet disparate need.

I commend *Cyclone Warriors* to you all.

General the Honourable Sir Peter Cosgrove AK AC (Mil) CVO MC (Retd)

BY THE SAME AUTHOR

- *The Sinking of HMAS Sydney* – how sailors, lived, fought and died in Australia's greatest naval disaster (Big Sky, 2023)
- *The Truth of War* - the realities of battlefield combat (Big Sky, 2023). Originally *Lethality in Combat*, (Big Sky 2013)
- *Bombers North* - a history of bombers operating out of Australia in WWII (Avonmore, 2023)
- *Attack on Sydney Harbour* – a commemoration of the 80th anniversary of the midget submarine raid on Sydney in 1942 (Big Sky, 2022)
- *Australia Remembers 4 – the Bombing of Darwin* – a book for young people, suitable for upper primary/lower secondary. (Big Sky, 2022)
- *Eagles over Darwin*: the USAAF defending northern Australia in 1942 (Avonmore, 2021)
- *Medieval Military Combat*: an analysis of battlefield techniques in the Wars of the Roses (Casemate, 2021)
- *Darwin Bombed*: the story for young people (Avonmore, 2020)
- *Atomic Salvation*: how the atomic bombs saved the lives of 30 million. (Big Sky, 2020)
- *Teddy Sheean VC*. Originally published as *Honour Denied*, Teddy Sheean: Tasmanian Hero, how an unfair system denied the Navy and Sheean a VC. (Avonmore, 2016) Republished following the award of the Cross in 2020. (Big Sky, 2021).
- *The Empire Strikes South*, an accounting of all Japanese air raids made in Northern Australia, showing the attacks were far more widespread than first thought. (Avonmore, 2017)
- *Carrier Attack*, (with Peter Ingman): an extensive technical analysis of the first Darwin raid, revealing many unknown aspects of that assault. (Avonmore, 2013)
- *The Submarine Six*: biographies of the six who had RAN submarines named after them. (Avonmore, 2010)
- *Darwin's Submarine I-124*, detailing the sinking of the first Japanese submarine by the Royal Australian Navy. Published by Tall Stories, 1995, as *Sensuikan I-124*, and by Avonmore, 2011.
- *By Derwent Divided* tells the story of the Tasman Bridge collapse. (Tall Stories, 1999, and three further editions.)

Out of print:

- *Zero Hour in Broome* (with Peter Ingman, Avonmore, 2010), analysed the second biggest air raid on Australia.
- *A War at Home*: the air raids on Darwin of the 19th February 1942 (Tall Stories, four editions, from 1999 to 2015)
- *Ten shipwrecks of the NT* (contributing author, Museums and Art Gallery of the NT, 2005)
- *Captain Hec Waller – a Memorial Book*. (Co-author, Edited by John Waller; Drawquick Printing, 2008)
- *Darwin Sayonara*, a novel for young adults. (Boolarong, 1990)
- *Wrecks in Darwin Waters*. (Turton and Armstrong, 1991)

EXPLANATORY NOTES

- Times of day are given where necessary as per sourced official documents. Military forces world-wide, both then and now, use a 24-hour system. Instead of having possible confusion over which of two possible "o'clock" is being referred to, 1pm in the afternoon is referred to as 1300. Half past one is 1330, and so on. So 6pm is 1800; 9pm 2100, and so on until 0000, which is midnight. Normally times are written with an "h" after each time; this has been routinely omitted here.

- Measurements using the Imperial system have not been modified.

- Ship speed, both in WWII and today, is measured in knots. This is the amount of nautical miles covered in an hour. A nautical mile is 1,852 meters, or 1.852 kilometres. In the English measurement system, a nautical mile is 1.1508 miles, or 6,076 feet.

- The person commanding the ship, no matter what their rank is referred to as "the captain". However, there is also a rank in navies of captain, which is the equivalent of an army's full colonel. A naval captain wears four rings on his sleeve. The "captain" of a ship though might be of any rank from warrant officer or above, or even in wartime extremes a petty officer – but still referred to by anyone on his ship as "the captain".

The confusing world of the rank system

Understanding the rank system of armed forces is both confusing and annoying. One of the most common frustrations is that armies have the rank of captain and so do navies – but in the first it is a "junior officer" rank, and in the second it is a senior designation. Air forces have "group captains." All three forces have lieutenants, but the ranks are not at the same hierarchical level. How does all this work? The table below provides some explanation.

Commissioned officers – hold the Monarch's Commission, which may be only withdrawn by an Act of Parliament		
Royal Australian Air Force	Army	Royal Australian Navy
Marshal of the RAF	Field Marshall	Admiral of the Fleet
Air Chief Marshal	General	Admiral
Air Marshal	Lieutenant-General	Vice Admiral
Air Vice-Marshal	Major-General	Rear Admiral
Air Commodore	Brigadier-General	Commodore
Group Captain	Colonel	Captain
Wing Commander	Lieutenant Colonel	Commander
Squadron Leader	Major	Lieutenant Commander
Flight Lieutenant	Captain	Lieutenant
Flying Officer	1st Lieutenant	Sub-Lieutenant
Pilot Officer	2nd Lieutenant	Midshipman

Warrant Officers – hold the Monarch's Warrant, and may not be summarily dismissed, nor subject to the punishment deemed suitable for the ranks below		
Royal Australian Air Force	Army	Royal Australian Navy
Warrant Officer	Warrant Officer Class 1 Warrant Officer Class 2	Warrant Officer
Non-Commissioned Ratings – may be dismissed within the system		
Royal Australian Air Force	Army	Royal Australian Navy
Flight Sergeant Sergeant Corporal Leading Aircraftman Aircraftman	Staff Sergeant Sergeant Corporal / Bombardier Lance Corporal Private	Chief Petty Officer Petty Officer Leading Seaman Able Seaman Seaman

A Note on Sources

The author began research on Cyclone Tracy in his own early days in the Navy, when he was tasked, given his civilian writings of military history books, with compiling a history of the RAN's efforts. Hence some sources are listed as being from those days, reflecting contact made with personnel who were present during and after the cyclone. The author also published *Wrecks in Darwin Waters* (Turton & Armstrong) in 1990, which was the first comprehensive examination of shipwrecks caused by Cyclone Tracy.

The original manuscript is heavily footnoted as to the origin of the information. These have been removed for reasons of space, with the exception of Chapter 10. Readers who wish to ascertain such evidence are welcome to contact the author through the publisher.

Some sections of this work have appeared in other publications by the author, especially in *The Truth of War* (Big Sky Publishing), *Wrecks in Darwin Waters* and in various popular and professional magazines.

MAPS

Darwin is remote from the main cities and population of Australia. Just prior to Cyclone Tracy hitting some civil and one military aircraft had flown to the relative safety of Katherine, 300 kilometres to the south of Darwin. However, the Christmas holiday period meant many pilots were not available leading to the loss of over two dozen aircraft at Darwin airport.

The day after the cyclone the first RAAF relief aircraft departed from RAAF Richmond, outside Sydney, and Canberra. These used Mount Isa in outback Queensland as a staging point. Subsequently RAAF relief efforts centred on the fleet of RAAF Hercules transports based at RAAF Richmond, some 3,000 kilometres from Darwin.

The Royal Australian Navy mounted Operation Navy Help Darwin with many ships being despatched from the main fleet base at Sydney. The key Australian Defence Force northern base at Townsville was also important in sending help to Darwin, and Army equipment later left this location via Navy ships and also overland.

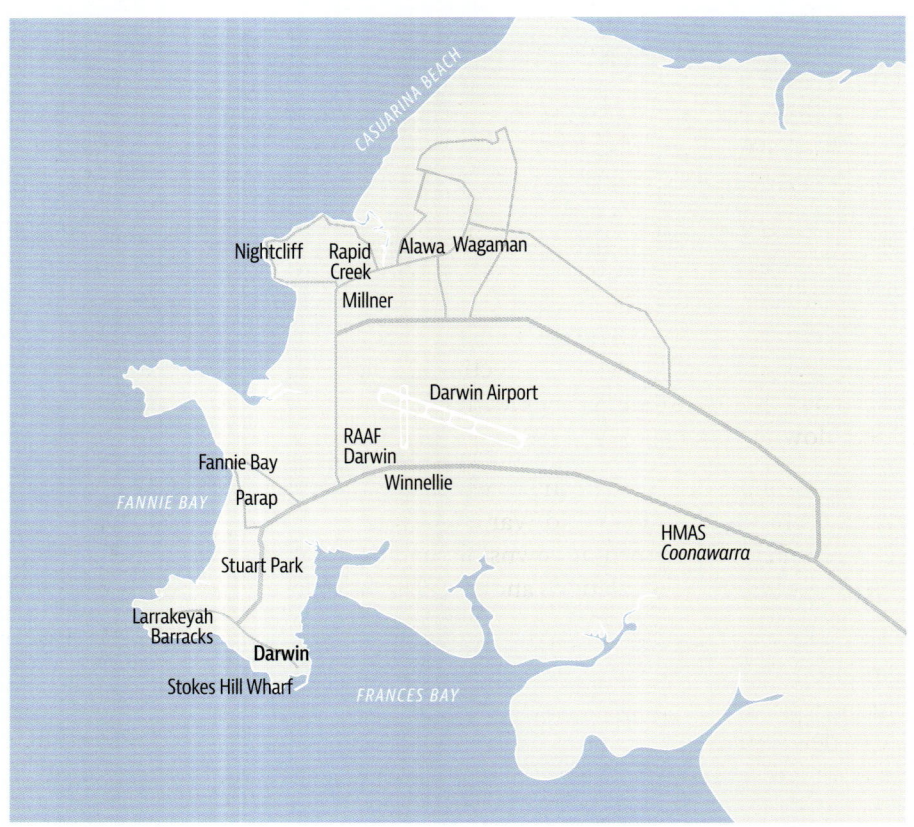

The city of Darwin was located adjacent to the Stokes Hill Wharf. From the 1930s a number of key defence establishments had been built, including nearby Larrakeyah Barracks and RAAF Darwin. Post-war RAAF Darwin also served as the civil airport, and by the 1970s a growing population had seen new suburbs built to the north including Rapid Creek, Nightcliff and Wagaman. While the impact of Cyclone Tracy was felt across the wider Darwin area, the northern suburbs were the hardest hit.

INTRODUCTION – CYCLONE TRACY AND THE ARMED FORCES

Cyclones are a fact of life in tropical northern Australia. People moving to the "Top End" can get quite alarmed as the cyclone season – basically November to April – approaches, and advertisements advising on what to do are heard on radio stations. They are advised to plan where they will shelter, check their house is strong enough; and to have a "cyclone kit" of tinned food, a radio, bottled fresh water, and so on. Most of them will have heard of Cyclone Tracy, which hit Darwin on 24 December 1974, and the widespread death and destruction it caused. All of these things can add up to a daunting prospect for the months ahead.

Tracy was an unusual cyclone. The writer has been through several small ones, and they were alarming enough.[1] To sit through one in a strong building is not too stressful, but it's more the actuality of not knowing what's going on outside – how many trees are down; how many roofs blown off, what will happen next – that adds to the tension. When the winds die down people can emerge to start inspecting the damage. In a big tropical city, there are always numerous trees that have been blown over, and quite often there have been power lines brought down by them.

Tracy was an exception, but then again Darwin was too in those days. The houses were often ramshackle, built hurriedly after World War II, without much attention to building standards. Post-war, the emphasis for returning townsfolk had been on establishing something to keep the rain off. There were few regulations and even less of a government inspection system.

Without the Australian Defence Force, Darwin would likely have even been abandoned as a city in 1974. With most of her population forcibly evacuated after the Christmas Day disaster, in the tropical wet season with torrential rain and high humidity; with no electricity, thousands of dead animals, and scores of dead humans, Darwin was a dangerous place to be, and survival would have been most difficult. But with the Army, Navy and Air Force to the rescue, the Northern Territory capital came back from disaster.

But almost nothing of the armed forces effort has been told. Most stories of Tracy centre around the thousands of people who had their lives wrecked. This book details the story of the cyclone and how some members of the services were directly affected, but then explains how the armed forces rallied for many months to the salvation of a city. Using archival photographs, location visits to what still remains, and interviews with ADF personnel, this account shows the nation's finest in some of their greatest hours, but also some of the untold stories behind the armed forces and Cyclone Tracy. It settles too, with extensive research, the number of fatalities which resulted from the cyclone, and dispels any stories of armed forces involvement in a cover-up. And it also argues for more formal recognition of the efforts and achievements of uniformed members involved.

Dr Tom Lewis
April 2024

---

1   The author has lived in Darwin for over two decades during three separate periods: 1968-1969, 1988-1999 and 2007-2021.

CHAPTER 1
THE ARMED FORCES IN DARWIN PRIOR TO TRACY

Darwin as a lonely outpost of the British Empire needed military forces from the moment government officialdom arrived by ship. The Royal Navy in the 1800s was what kept the Empire secure, and warships and Marines – seagoing soldiers – were part of the settlements as various sites were sought for a colony. The small peninsula where Darwin's location was eventually decided upon was well inside the huge harbour, and from those early times the military was a presence.

World War I had little impact on the Northern Territory but the port being used as a transit point quickly became a usual practice. The Navy's first submarines, *AE1* and *AE2*, staged through it going south, and refueling ships going north were commonplace. Lord Kitchener, the mustachioed hero of the British Army and eventually a household name due to his presence on recruiting posters, arrived on a visit and pointed out the need for guns to defend the harbour. These and more vital infrastructure – oil tanks, an aircraft runway, and the telegraph line connecting Australia with the world – all became part of the northern presence and necessarily needed defending.

World War II had a tremendous military impact on the Territory, and in many ways it was the centre of military operations within the country. Certainly, by the marker of enemy attention it was: Darwin itself saw over 60 air raids, out of 207 enemy missions across the country's north.[1] By the time the enemy arrived the Navy had a strong presence. A wireless signal station named *Coonawarra* had been set up inland, and HMAS *Melville*, a shore base that grew steadily in size, was located in the town itself. Out on the harbour a myriad of small craft and a few bigger ships called the port home. One of them, the corvette HMAS *Deloraine*, sank the RAN's first Japanese submarine on 20 January 1942 – 80 men remain inside the intact wreck of the *I-124* outside the harbour to this day.

The ensuing air raids however took a steady toll of the Navy's ships and lives. Many vessels left the port never to return, one of them being HMAS *Armidale*, carrying the Navy's first Victoria Cross recipient to his death – Teddy Sheean fought valiantly at his anti-aircraft gun as the corvette sank beneath him and 101 shipmates died.[2]

HMAS *Patricia Cam* was bombed by a Japanese floatplane which then landed and captured Reverend Len Kentish. He was never seen again, and his fate was unknown until post-war when it emerged he had been executed in an angry fit of revenge by a Japanese officer. It was a war fought at extreme distance and savagery for the Navy.

The Army by comparison had a more localised time in Darwin and surrounds but it was by no means an easy war. Added to the extreme heat were the frequent raids by enemy bombers, sometimes several score at once. The Army fought them with anti-aircraft guns, ringing the enemy's intended targets as far south as Katherine, 300 kilometres inland but still bombed. When the Japanese switched to night raids the Army brought in searchlights. Eventually radar was added to the mix together with radios to talk to the aircraft taking off to do battle. There

---

1   Compiled in the same author's *The Empire Strikes South*.

2   See the same author's *Darwin's Submarine I-124*, and *Teddy Sheean VC*.

were thousands of troops based in the north; even on the day of the first raid Jack Mulholland, an Army anti-aircraft gunner, estimated some 8,000 in town.

The Army also supplied the bulk of infrastructure management, including operating the centres under martial law following the move of the administrator and his seat of governance to a safer Alice Springs. Field hospitals sprang up, as did the vegetables and fruit grown in a myriad of military-supervised gardens, many employing Aboriginals evacuated their northern homes. *Army News* became the only newspaper in town. One let-down for the Army though was that they never got to use their huge 9.2-inch coastal defence guns in anger: the Japanese only came by air.

When the enemy first arrived on that fateful Thursday morning on 19 February 1942 the only air defence was ten P-40 fighters flown by the United States Army Air Force. The Royal Australian Air Force had been expanding but mostly as a training organisation supplying aircrew to fight in Europe. The USAAF fought hard for most of the rest of the year – the four pilots who lost their lives on that first day were joined by plenty of other airmen. By then the RAAF, and incidentally the RAF too, began arriving with Spitfires, and the air above northern Australia saw many a dogfight and many an aircraft plunge smoking into the sea. The Japanese lost 62 machines all up, and most of them went down with their crews.

By 1943 Allied bombers had begun moving in. It was time to turn the tables around: instead of receiving heavy ordnance from the skies Allied squadrons would start delivering it instead. The Americans also brought with them what wins war: logistics. They built roads – finishing the Stuart Highway as part of it – and airstrips, hardstands, taxiways, and hangars, some of which can still be seen today. By the time the war ended there were 51 airstrips in the Territory, and several of them were home to USAAF bombers. The third force though, most unusually, was a Dutch squadron which arrived ready to seek revenge for the price they had paid when they were sent packing out of the Netherlands East Indies – many of Broome's 86 air raid casualties from March 1942 were women and children from Holland.[3]

As the war moved north the forces went with it, and the Territory became less of a massed military presence. Many of the families who returned north found their homes either demolished or occupied by the military, but gradually civilisation returned. The military stayed though in smaller numbers. The US Army Air Force never really went away – they just saw a slight name change to the United States Air Force, as America followed other countries in realising a third separate force was the best solution, rather than wings of the navy and army.

The Korean War saw both Australians and Americans staging north through Darwin, as they did during the Vietnam conflict. As the Cold War grew so did the USA's presence: many residents recall "supervising sunset with a coldie" and seeing the giant eight-engined B-52 bombers carrying their nuclear deterrent payload take off into the northern skies.

Meanwhile the armed forces presence grew to become the eyes of protecting the northern border. By the time Tracy arrived the Navy had a permanent patrol boat presence as well as several long-distance communication bases. The RAAF still owned Darwin's airport (which had the longest runway in Australia), and uniquely occupied it and controlled civilian aircraft landing upon it. And the Army's wartime guerrilla-style Nackeroos developed into the post-war equivalent of NORFORCE which still operates today.

With its rich military heritage, Darwin was very much the home for the armed forces in 1974.

3   See the same author's *Bombers North*, and *Eagles over Darwin*.

Central Darwin in the 1970s, with Stokes Hill Wharf visible at the top.

Typical post-war housing in Darwin's outer suburbs. Built quickly with the main living quarters usually on the first level for tropical comfort, such housing was devastated by the cyclone.

CHAPTER 2
THE CYCLONE STRIKES – THE AUSTRALIAN DEFENCE FORCE STORY

The cyclone that neared Darwin in the week before Christmas 1974 was small but gathering intensity as it approached. The diameter of its gale force winds was only about 100 kilometres, in contrast to some north Pacific typhoons, which have had diameters of 1,500 kilometres.

The capital of the Northern Territory was a big sprawling town with a very strange layout. It was centred around its huge airport, with Australia's then-longest runway taking up a 3,354-metre length. The suburbs had started off on a small peninsula in the south – much bombed in WWII – and by 1974 had advanced up the left-hand side of the airport complex and were creeping across the top side. In January of that year Darwin's population had been counted as 46,656.

Darwin was no stranger to destruction: it had been heavily damaged by cyclones in 1897 and 1937, and as already noted, WWII had impacted the city heavily. To make the oncoming storm impact worse, the city buildings were not ready for destructive winds. This was not a topic of discussion for the population on Christmas Eve, but it was very much on their minds over the next weeks and months. While government centres were built strongly – the schools designated as cyclone shelters stood up to Tracy very well – the housing building codes were by no means adequate. Following the cessation of military rule post-war, the emphasis had been on building quickly. Even the government-built houses – all aligned to the same angle in rows in the newer suburbs – were by no means strongly built. As a result, the damage to housing would be extremely destructive.

Cyclone Tracy had been named on 21 December, but there was little notice of it taken as the local residents busied themselves with Christmas preparations. The Top End was a hard-drinking and partying society, and this along with Easter was one of the two biggest holiday periods of the year. It was now the hottest and most humid time of year too, and cold beer was the antidote. Last minute shopping and an atmosphere of celebration was usual, and any thoughts of danger were not present in most minds. To make matters worse there had been destructive wind warnings in previous weeks, but Cyclone Selma had hardly impacted Darwin in early December.

The first warning that Darwin itself was threatened came at 1230 on 24 December, approximately twelve hours before the onset of destructive winds. By the early evening it was apparent the cyclone would impact the city, although whether it would pass directly over the built-up areas was not that clear. In the event it did, with a recorded surface wind speed of 217 km/h.

Darwin at the time was not a large city, perhaps about 20 kilometres across. As a result of the expansion there was little of the built-up areas that was not in the path of the cyclone, which took a track from north-west to south-east across the houses and shopping areas.

The experience for armed forces members and their families was little different to those of the civilians of Darwin, that is, unforgettable and terrifying, often with injuries and sometimes death. Many families had gone on leave, either locally or "down south", taking off in previous

RAAF C-47 A65-68 seen at Bankstown, Sydney, in September 1974 just three months before Cyclone Tracy. On the afternoon of 24 December 1974, it was flown from Darwin to the relative safety of Katherine, 300 kilometres to the south. (Nigel Daw)

days to see family and friends. However, there were operational structures still in place. For the Air Force, for example, commanding Group Captain D Hitchins was on leave some 130 miles away; but in his absence Wing Commander WJ Monaghan was in charge. This is always the organisation of the forces – a clear and defined command structure is in place.

The RAAF did the best it could to safeguard its aircraft. One C-47 transport was flown out, but there was no pilot for another, and the only helicopter pilot who could fly an Iroquois at the base was medically unfit to fly.

Continuing with the Air Force, by 2143 a bus had transported families from "igloo type quarters, the safety of which, under extreme weather conditions, was suspect, to the airmen's block which was a reasonably substantial structure. By 1025 the RAAF guard dogs had been moved to the detention cells"; and "Father Grannall, an RAAF padre [priest] was advised to cancel midnight mass."

Around 2200 the RAAF requested the presence of an Army medical officer at their base. A Captain Strickland was despatched from Larrakeyah Barracks and arrived at the Air Force base around 2300. Later, around 0230 on Christmas Day, a Corporal Purcell made his way on the reverse journey, from the RAAF to Larrakeyah Barracks, to obtain a radio to keep communications going between the two forces.

Then Flight Sergeant Ken Stone and his wife Margaret remember going to a Christmas party before returning to their married quarters on the RAAF base. As the winds increased, they had their youngest child, then one month old, in a papoose arrangement in front of mum; and "tied the other two kids around our waist – one around mine and the other around Margaret's." Then the roof of the house blew off, and in one of those later-comic moments, Margaret tried to save the new and expensive stereogram. Down below the collapsing house they took shelter under a fallen tree, and then threw items out of the garden shed before sheltering inside. When the eye of the storm passed over the two parents went into the house and recovered useful items – Ken recalled "Margaret emptied out the washing basket and put Christmas presents in it."

Lorraine Dixon, an Army wife, had been in town since July 1973. Her husband Greg was in Signals, based at Larrakeyah Barracks, where they also both lived in a service house at 66 Clowes Street.

Lorraine was working as a Registered Nurse in the NT Government's Home Nursing Service in Parap. She was three months pregnant when the cyclone arrived. On Christmas Eve she was at work, and was at Fannie Bay Gaol, as the Nursing Service supplied a nurse to work there. She remembers seeing prison officer "Happy" Hampton, who was to be one of the 66 fatalities, as the day progressed.

Lorraine went home and together with her husband closed things as much as they could for the cyclone, and then went to the neighbour's next door for some Christmas cheer. She remembers:

> When the winds got up we were inside, and then the alarm went off for the base and my husband and my neighbour's husband went off to their Army place of duty. I remember looking outside and could see a power line tripping and sparking in the wind, and then the power got turned off. My husband was later allowed to come home. There was someone assigned in each street to have a telephone and a flashlight. We went to bed but I didn't go to sleep – the whole house was creaking and groaning. I went to the loo a lot more out of nervousness than being pregnant. I remember the water in the toilet bowl was draining away and then coming back with a big gurgle.

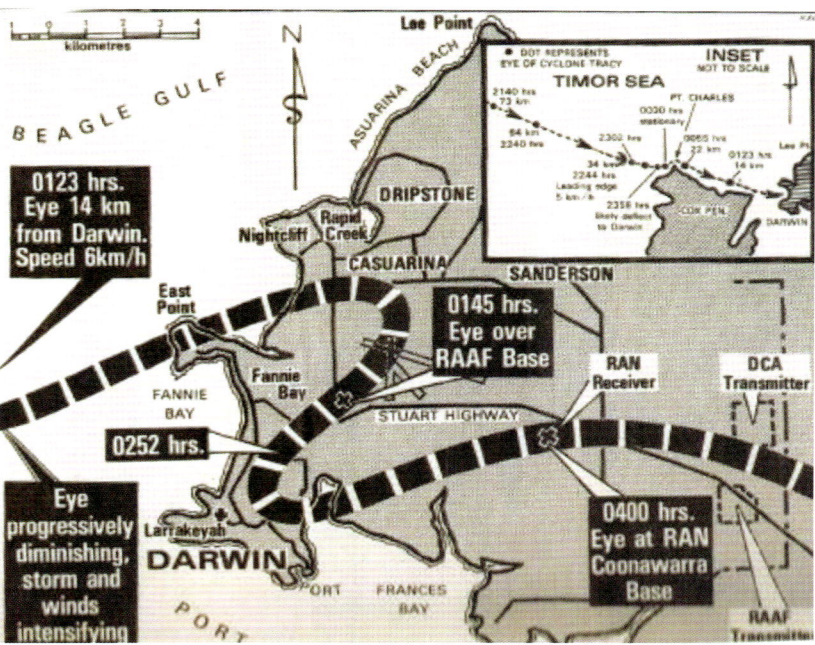

A newspaper graphic showing the track of the cyclone. At 0145 the eye was over the RAAF base and at 0400 it was over the naval communications facility HMAS Coonawarra.

The walls were going in and out, and the louvres gave way – my husband was cut by the glass but not seriously. We took shelter in the bathroom as they told you to do. My husband Greg shone our torch out and you could see the ceiling lifting off the walls. He said we've got to get out of here. We had a little dog. We went to the neighbour's but they did not hear us, and then we saw our house being swept away as if by a giant hand. We managed to get into the neighbour's Besser brick storage unit where they had two mattresses. We lay on one and put the other on top of us until daybreak.

Lorraine assisted at the Army Medical Centre and stayed for the next two nights with her husband at the Army Signals centre. An offer was made for her evacuation, and as she was pregnant, she made the decision to leave. She left for Sydney on a C-130 evacuation flight, supervising the two boys from their neighbour's, and she was allowed to bring along their dog too.

At the Navy base the then-Executive Officer, Lieutenant Frank Densten, had one of the more notable stories from the night of the cyclone: he had seen an aluminium 15-foot boat blown through the air above him when outside. Yvonne Lowe, a senior WRAN, remembers that:

… we were sheltering onboard wherever we could find a safe area. The majority of the ship's company ended up in the Junior Sailors Mess. Just prior to sunrise, women and children sheltered under tables when the roof threatened to collapse on us.

Families far and wide, inside the base and out, found their houses disintegrating. A young boy who would later become a Navy member and then Chief of the Defence Force was one of them. David Johnston, who in 2024 as a Vice Admiral assumed the highest post of all, spent four years in Darwin as a child and went through Cyclone Tracy before his family was evacuated. His father led the Northern Territory Tourist Bureau, which became untenable after Tracy, and so the family headed south.

Meanwhile the Navy Headquarters in Darwin's CBD were disintegrating under the winds. The RAN maintained a duty radio watch there, in contact with their ships and with bases in the south of the country. The Naval Officer Commanding Northern Australia, Captain Eric Johnston RAN, later said:

At 25 minutes past midnight, I made my final telephone call to Navy Office, Canberra, advising the duty staff officer of Tracy's imminent presence in the city and giving my prediction that severe damage would occur right throughout the city. At this time heavy rain was falling, the trees opposite Naval Headquarters, which are now my[1] Administrator's offices, had all been uprooted, a car passed down the Esplanade some twenty feet in the air and the end of the building in which I was located began to disintegrate. I well recall the final words from the naval duty staff officer in Canberra which seemed somewhat fatuous although well meaning, wishing happy Christmas to myself and my staff.

Five minutes later all phone and radio contact was lost and by 0430 the building had been destroyed and my three staff and I were buried. Three of us managed to dig ourselves out and take refuge in the centre cell which also acted as the Naval Headquarters bar. To my intense delight, not only did I have a packet of cigarettes which were not touched by

---

water, but the cell bar fridge, although without power, contained some still cold beer. I say without equivocation that the cigarettes and two beers I consumed are the best I have tasted in my life.

But of all of the Armed Forces the Navy had the most problems. They had four vessels in port, and in a practise followed by most mariners who could man their vessels, they put to sea in an effort to get away from the large immoveable structures which would certainly smash them – wharves, jetties and the like. But the tactic was not to work completely.

The official chain of events concerning the Navy's ships was:

- Patrol boat commanders met with NOCNA at 1400; ordered to sail at 1600.

- *Arrow, Assail* and *Attack* secured to buoys; *Advance* "anchored nearby".

- Around 0025 on Christmas Day NOCNA lost contact with Canberra when "his Operations Room disintegrated".

- *Assail* cable parted at 0158; left harbour; returned 1300 on Christmas Day.

- *Advance* weighed anchor at 0130; left harbour; secured 1200.

- *Attack* could not slip; cable parted 0125 and went aground in Doctor's Gully.

- *Arrow* broke her mooring at 0245; CO decided to beach in Frances Bay on mud; radar failed; struck Stokes Hill wharf bow first just before 0400. Abandon Ship ordered.

The loss of a small warship

HMAS *Arrow* had been in service since July 1968. One of twenty Attack Class patrol boats ordered for the RAN in November 1965, the ships' primary role was to conduct patrol work in Australian territorial waters. She had been home ported in Melbourne where her primary role was to conduct reserve training. On 30 July 1974, *Arrow*'s time as a RANR training vessel came to an end and she resumed duties with the RAN fleet based at HMAS *Waterhen* in Sydney. Following a period of work-up and trials, she departed Sydney on 21 August for her new home port of Darwin where she joined the Third Australian Patrol Boat Squadron. She arrived in Darwin on 2 September.

The following is taken from the Navy's Official History, which used extracts from the Board of Inquiry later held into *Arrow*'s loss:

> Her first operational tasking was to conduct a hydrographic survey of Cone Bay and Collier Bay in northeastern Western Australia. She arrived back in Darwin on 9 October. She sailed again for her first Fishery Surveillance Patrol off the Western Australian coast on 23 October and maintained a regular patrol and maintenance program up until what would prove to be a fateful Christmas period.
>
> As Tracy approached Darwin on the 24th the sailors of the patrol boat force were recalled from leave. Married members were instructed to assist their families prepare for the cyclone and then return on board in time to sail. The intention was for the four patrol boats to ride out the cyclone either at the moorings or at anchor in the harbour, and *Arrow* indeed had secured to the furthermost south-east buoy by 1830 that evening. By dusk, all of the patrol boats had been secured to their respective Commanding Officers' satisfaction.

The patrol boat HMAS Arrow in 1974 shortly before deploying for Darwin. (RAN)

As the clock passed midnight into the morning of Christmas Day, *Arrow*'s Commanding Officer, Lieutenant Robert Dagworthy, RAN, felt his ship was riding the storm well. Some 20 vessels were shown on radar astern of *Arrow* in the harbour at midnight. Over the following hour to 0100, the weather deteriorated, and Lieutenant Dagworthy felt that the cyclone was closer than had been predicted. The weather continued to deteriorate to 0200 and the number of vessels visible on radar had noticeably decreased. By this time the rain and sea spray were driving horizontally, and the visibility was virtually nil. The wave height could not be estimated due to the poor visibility.

In spite of the conditions, *Arrow* was riding the storm fairly well but there was some concern that the shackle used to secure to the buoy might part. Meanwhile it was noticed that the other three patrol boats had all disappeared and it was thought that they had gotten underway having dragged. By 0230 it was hoped that conditions were starting to improve; however, at 0245, Lieutenant Dagworthy was informed that the gypsy on the mooring winch had failed, and the ship was no longer secured to the buoy.[2]

Arrow's main engines were started with the intention of riding out the rest of the cyclone where she was. She was turned into the wind with satisfactory results, but occasional wind gusts caused the ship to roll violently. It was only with outstanding ship handling that *Arrow* was able to maintain her heading.

Conditions worsened yet again after 0300 and sometime between 0330 and 0345, the starboard engine alarm sounded, and it was ordered to neutral. The engine had lost circulating water suction and its temperature was reading 215°C. It could be used if it was essential, but only for short periods. Lieutenant Dagworthy was aware that both engine circulating pumps took their suction from the same manifold and felt that it was only a matter of time before the port engine would suffer the same problem. He favoured the port engine thereafter by necessity, but he assessed that the ship could not survive indefinitely on one engine.

The ship was being pushed westward and Lieutenant Dagworthy decided to beach *Arrow*

---

2   The gypsy is a toothed essential component of an anchoring system, responsible for controlling the chain or rope that connects the anchor to the vessel.

in Frances Bay where there was plenty of mud, no rocks and some measure of protection. His only concern by this stage was the safety of his crew and, if possible, to minimise the damage to his ship. All spare hands were ordered to the Flying Bridge with lifejackets partially inflated.

With the starboard engine overheating it was decided to use the port engine to turn the ship to starboard but it was not long before the port engine alarm also sounded. It too was overheating but with the safety of the ship and her crew now in peril, the throttle was not altered. With the ship turning to starboard, Stokes Hill Wharf unexpectedly became visible ahead and a collision was unavoidable. Just before 0400, *Arrow* struck Stokes Hill Wharf bow first and "abandon ship" was ordered.

Apart from the few crew members closed up in the wheelhouse, the ship's company had mustered on the Flying Bridge. The Executive Officer, Sub-Lieutenant John Jacobi, searched along the ship's port side for the safest escape route but, in attempting to reach the wharf, fell into the sea. He was washed ashore below the wharf gatekeeper's hut.

An article later published by the Department of Veterans' Affairs article added: "Some of the men were injured and in a very bad condition. Kevin Rainbow, the electrician, had serious lacerations and the Navigating Officer, Sub-Lieutenant Andrew Birtchnell, was also found suffering from hypothermia, and hospitalised."

Most of the rest of the crew scrambled on to the wharf by climbing onto the funnel casing, then on to the portside guardrail and on to the wharf safety rail. In addition to Sub-Lieutenant Jacobi, three others, including Lieutenant Dagworthy, ended up in the sea either by choice or misfortune but were eventually washed ashore. Those that escaped onto the wharf became separated in the poor conditions but were later reunited ashore where they sought shelter after getting off the wharf. Several suffered various injuries from flying debris.

Tragically two crew members, Petty Officer Leslie Catton and Able Seaman Ian Rennie, lost their lives as *Arrow* was wrecked under the wharf. In the confusion their movements following the order to abandon ship are unclear but at some stage, both fell into the water and drowned. It appears that at least Petty Officer Catton made it onto the wharf but was either blown off by the strong wind or was knocked off by flying debris.

In a memorable phrase later one of the *Arrow* survivors said that giant steel shipping containers were being blown about "like children's blocks".

A Board of Inquiry was convened to investigate the loss of HMAS *Arrow* and the damage to the other patrol boats. In *Arrow*'s case, the Board stated in its report; "The orderly action of the ship's company and the mutual assistance given in the prevailing conditions, firstly to reach the wharf and subsequently to leave the wharf area, can only be regarded as most commendable. Having regard to the position of the ship and the conditions this was obviously a major factor in the majority of the ship's company reaching safety." Able Seaman Robert McLeod was awarded the Australian Bravery Medal on 16 September 1977 for assisting injured shipmates.

Petty Officer Catton and Able Seaman Rennie are commemorated at HMAS *Coonawarra* with a plaque and the inclusion of their names in a stained-glass window at the base's chapel, and the *Arrow* Bar at the base is named in the vessel's honour. Arrow Drive, Catton Court and Rennie Road are all also named in their honour.

In *Navy News* of 17 January 1975, it was described how, in an effort to counteract the storm, the naval crews paid out massive amounts of steel cable to the naval cyclone buoys, which were specially designed to be used to combat cyclonic winds. The vessels used their engines to remain head on to the storm. The tactic was quite successful, although Lieutenant Paul de Graaf said HMAS *Attack* was "dragging the buoy after us" as his ship was thrown around the harbour in mountainous seas. As gyro compasses and radar sets failed, stray fishing trawlers and smaller boats had to be avoided by visual means.

The patrol boats had been built to tough specifications. Lieutenant Chris Cleveland, Commanding Officer of *Assail*, commented wryly to the *Navy News* that: "we proved the main engines don't fall off the mounts past 72 degrees of roll, even if the battery charger does." *Assail* actually managed to reach 80 degrees of roll and at one stage immersed her side navigation lights in the water.

The later Lieutenant Commander Paul Blanch, then the Executive Officer of *Attack*, remembers how his ship survived the storm but ended up beached on the rocks near Larrakeyah Barracks. But while the ship was high above the waterline as Christmas Day dawned she was certainly not dry. "The amount of water pouring off the cliffs and down onto the patrol boat was phenomenal", he recalls. *Attack* was later refloated by Navy divers and spent some considerable time being repaired. Paul Blanch, meanwhile, was also being repaired: he had been injured with a broken foot during the ship's rough ride.

Interstate newspapers reported initially that ammunition on board the *Arrow* blew up, but this was found later to be incorrect. Captain Johnston was reported in the January 1975

HMAS Attack survived Cyclone Tracy but ended up beached on the rocks near Larrakeyah Barracks. (RAN)

edition of *Navy News* as saying that "*… Arrow* did not explode, it hit the wharf and sank".

Another strange story concerned the captain, Lieutenant Dagworthy. He was reported in *Warning: the story of Cyclone Tracy* in 2014 as "found alive, floating in his life raft, some thirteen hours after he took to sea in it." When consulted later about this, Bob Dagworthy said:

> … that caused me great upset. At the time she was writing her book my name and I were easily accessible. Also, if she had bothered to read the Board of Inquiry report she might have got the facts right. Our life rafts with the hydrostatic releases were washed away early in the night. I like all of my crew was wearing a life jacket and that is what saved my life.

In fact, Lieutenant Dagworthy's leaving his ship was in the finest traditions of naval services, in that he was the last man off the vessel, as the Board of Inquiry noted:

> He then went below to the lifejacket locker in the wardroom flat and found it empty, returned to the Flying Bridge and went aft to find that everyone had apparently left the ship.

> The captain then left the ship himself. By now the quarterdeck was awash and believing that he could not get directly ashore he jumped into the sea over the starboard side and was swept through the [wharf] piles receiving lacerations en route. He was finally washed onto the foot of the sea-wall adjacent to the wharf gatekeeper's hut.

Captain Johnston noted a further special and tragic loss of life impacting the RAN:

> To add to the Navy's losses, four dependents, two wives and two children were also killed in their married quarters while their husbands were at their place of duty, and thus the Navy in Darwin with some 1.5 per cent of the population suffered 12 per cent of the total fatal casualties and indeed was the only service to so suffer.

The sailor who lost a wife and two children was Able Seaman Geoffrey Stephenson. He later flew to see his wife's parents in Sydney, and then worked briefly on relief operations at HMAS *Kuttabul* before returning to Darwin.

The eye of the storm crossed in the early hours of Christmas Day. Two reliable reports of the actual period of the calm were obtained from Fannie Bay and the RAAF log at Darwin airport. The period of "complete calm" at Fannie Bay was from 0320 until 0355, while at the airport it extended from 0350 until 0425.

RAAF sergeant photographer Ken Markwell had survived the night with his family of four children and his wife. He remembers that the roof of the house went; and they all sheltered in the bathroom. He recalls when the cyclone finished he went outside, and later said: "There were no leaves on any tree; you could see right through Darwin. The silence of the place was terrible; there were no birds – nothing." His family was evacuated two days later to Brisbane and then to Mackay to Ken's parents home.

By the time the residents of Darwin emerged from their places of shelter it was apparent massive damage had been done. The damage on the harbour was extensive. According to the harbourmaster's report: "At least 29 vessels were sunk or wrecked". These included several of the large prawn trawlers that were based in the port, a passenger ferry, the *Darwin Princess*, and the largest vessel in port, the steel three-masted schooner *Booya*. Neither of the latter two were seen again for decades: *Booya*, who took five people with her, was found in 2003, and *Darwin Princess*, who went down with her skipper Ray Curtain, was located in 2004.

The Darwin suburb of Wagaman after Tracy's impact. (Bureau of Meteorology)

Another loss, which was strangely related to the Navy, was the Fairmile launch *Ataluma*, which was one of 35 such ex-RAN vessels from WWII. She had worked as a whale-chaser post-war, and was then repurposed as a survey ship, but had just been bought by Steve Pastrikos of the Aspa City Hotel. Torn from her moorings, she had been swept unmanned down the harbour to East Point where she was driven ashore. The wreck of this once proud wooden ship – likely Fairmile *ML 808* of the Navy – can be seen scattered in pieces near Dudley Point to this day.[3]

On land it was not immediately apparent the worst was over. At Navy base HMAS *Coonawarra* Yvonne Lowe remembers that morning:

> … drying any bedding we could find and dragging any wayward roofing iron into the tennis courts to prevent it from doing further damage should another blow occur. There were strong rumours another blow was on the way.

The extent of the damage is best expressed in raw statistics:

- Darwin Naval Headquarters was destroyed, as was 80 per cent of the patrol boat base and 90 per cent of the naval married quarters.

- The oil fuel installation and the naval communications station HMAS *Coonawarra* were extensively damaged.

Further afield, the scale of the damage statistics shows that life would be almost impossible for many of the city's residents:

- Approximately 20,000 families had their houses impacted in a way that meant they were not able to be lived in. That meant that 90% of housing was unliveable. Grant Tambling,

---

3   The author wrote "What Fairmile is that?", an article which has been released in several publications, analysing which of the Fairmile fleet the *Ataluma* was. The nearest conclusion arrived at so far narrows it down to *ML 808* but the precise identity remains uncertain due to lack of records.

then a member of the Legislative Assembly and later a Senator for the NT, who went through the cyclone with his family, later said: "I believe we were among the 10% of the population to retain their house."

- 650 people were injured and treated in Darwin (many more were attended to interstate).

- Initially all external and internal communications were destroyed. In 1975 there were no mobile phones, email or the internet. Houses with telephones connected to an exchange – if the exchange was damaged then there was no telephone service.

- Water, sewerage, and power services had been contaminated, cut or severed.

However little comprehension for those imagining Tracy's impact can be gleaned from words. Instead, the photographs taken at the time – most of them by Defence – show the true reality of the shattered capital city of the north.

Cars in the swimming pool of the Travelodge Hotel in Darwin, circa Christmas Day 1974.

Flattened houses in one of Darwin's outer suburbs.

Part of Darwin's main business district showing the extensive damage from Tracy. This picture was taken a few days afterwards as the streets have been cleared and vehicles are driving about.

A12-125, one of two BAC-111 RAAF VIP jets operated by No. 34 Squadron based at Fairbairn, Canberra, in the 1970s. On Christmas Day 1974 this aircraft flew Major General Stretton from Canberra to Mount Isa, where he boarded a C-130 for the flight to Darwin. It is seen in Sydney in 1972. (Nigel Daw)

Alan Stretton (right foreground) with Prime Minister Gough Whitlam (centre), during the prime minister's visit to Darwin in the week after Cyclone Tracy. Note the casual clothes worn by Stretton. (ADF)

CHAPTER 3
THE CONCEPT OF RESCUE, AND MILITARY COMMAND

In no sense, despite occasional reports to the contrary, was "martial law" declared in Darwin in the days following Tracy. Martial law would mean legally the armed forces would be in control, and have the power to arrest and gaol people, and run matters in the way they chose. That did not happen. But in an undeniable way the immediate response was led by an Army man, and his rule for the short time he was there was absolute.

Six months previously, Major General Alan Stretton was still an Army officer when he had taken over the newly formed civilian Natural Disasters Organisation. It was a tiny affair, only fifteen people strong, and by the end of the year had just moved into its new premises in Canberra. The structure of the new organisation had been agreed between the federal government and the states. Its main task, in Stretton's words, was "co-ordinating Commonwealth resources in the event of a major natural disaster". Within months it had a test of its new abilities when a large bushfire, measuring 900 kilometres around, was out of control in southwest New South Wales.

The very concept of what was to happen in the ravaged city of Darwin was a curious one. Australia, like the United States, but unlike the United Kingdom, is organised on a "states and territories" basis. Each has its own parliament, some of them being two houses, and others unicameral. They function under the Westminster system, where each has a Governor (in the case of a territory an Administrator), who is appointed by the government of the day, but who has oversight of the government. In the Northern Territory's case in 1974 it had an Administrator, and a form of government, albeit one of local councils and federal government departments. Self-government had been announced in 1974 but it did not occur until 1978. So why did the Territory simply not continue running affairs as before?

In the case of a stricken Darwin from Christmas Day onward it was rather different than if a cyclone had smashed, for example Brisbane. The population of the northern capital was tiny by comparison with the southern cities – 40,000 or so compared to millions. Many of Darwin's population would be incapacitated, either physically or by the impact on their families. Resources in the north were limited; everything would be available in smaller quantities: from bandages to fuel to food to clothing, with much of those resources, it was presumed, destroyed by the cyclone. Many of these thoughts would have doubtless been going through the heads of those in some sort of command who were either in Darwin, or those down south whose attention turned northwards. The fact it was now Christmas Day made matters so much worse, as almost the entire country was in a closed-down situation. In 1974 there was no instantaneous news service via the internet, nor was there as effective a communications network. Stretton, it turned out, was going to be the right man in the right place at the right time.

To give him the full title he later retired with, Major General Alan Bishop Stretton, AO, CBE, a graduate of the Duntroon Army College, had served in four conflicts. Born in 1922, he commanded an infantry platoon in WWII, was deployed again in the Korean War and the Malayan Emergency and served in Vietnam. He was a resourceful man: he had played successfully in major football competitions, and in his own time he had pursued the study of law and had been admitted to the Bar in 1969.

However, it took some hard thinking to make a national "command decision" as to whether the federal government should override the Territory's. Stretton wrote later:

> ... was it a minor event that could be handled locally or were we confronted with a major disaster with perhaps thousands of lives at stake?

He hastily organised the NDO, and then requested an RAAF aircraft from the Minister of Defence. *En route*, it was decided that he would be, in the words of the Acting Prime Minister, Dr Jim Cairns:

> ... the supreme authority. His orders will only be countermanded by the Prime Minister or myself.

A 1959 portrait of Jock Nelson in his role as a Member of the House of Representatives, before he became Administrator of the Northern Territory.

Contrary to some reports, Stretton – nor any other senior ADF person – did not order the entire defence force to be mobilised: "... every member of the Australian Defence Force was recalled from holiday leave" as *Woman's Day* put it. Rather, he focused on the task immediately to hand, and the rest of the services, in many respects, simply joined in.

By the time he landed the Administrator had been sidelined and Stretton was now the legal and effective commander of the city, although as he later wrote, it would be a combination of military and civil:

> I had no regrets at deciding that the armed forces would not take over the City of Darwin but would provide the backup and support to the local population who would do the job themselves.

For all that, Stretton was not in Army uniform. He wore no badges of rank and was only accompanied by one aide – Major Frank Thorogood – rather than the usual numerous staff a major general would be using in a time of war.

Installing Stretton was a curious decision in some ways. The Territory had the near equivalent of a governor: an Administrator, John "Jock" Nelson, but he was sidelined by the arrival of a military commander even though martial law was not declared. Sophie Cunningham suggests that "in a practical sense there was little he could do in a crisis like this". But apart from being trained and capable in scenes that resembled the aftermath of battle, was there a need for Stretton?

One explanation is that the concept of something like Tracy had been explored at a national level:

> The National Disaster Organisation was formed in August 1974 with only a small staff of fifteen (Stretton 1975) and had little experience dealing with large-scale disasters (Britton & Wettenhall 1990) and was still exploring its mandate (Emergency Management Australia 2005). Cyclone Tracy was the organisation's "baptism of fire" (Jones 2019). The organisation's role was to coordinate national efforts with other state-based and voluntary

agencies during major natural disasters or other civil emergencies (Jones 2019). The National Emergency Operations Centre was opened and exercised for the first time in October 1974 (Jones 2010).

Nelson was no slight figure to be dismissed so summarily. The *Northern Territory Dictionary of Biography* notes his military service, cattle station ownership, service in politics in both local, the NT's Legislative Council, and then in federal representation. Nelson was the first Territorian to hold the post of Administrator, a territory's equivalent of a state's governor.

In any event, the Administrator was sidelined, in a rather major way. Stretton did not formally meet him, although Nelson was present at the first, and subsequent meetings of the co-ordinating committee which met to govern Darwin. When Stretton's aircraft landed, he had an initial meeting at the RAAF operations room at the airport and was then driven into town in a Kombi van, with some difficulty through the wreckage, to a hastily convened meeting place in the police station. Immediately decisions were made and men – there were no women in the organising group – went about their orders. Indeed, Stretton later wrote, it was not "until some days later that I found out that "Jock" was the Administrator of the Northern Territory, Mr JN Nelson."

Indeed everyone was, as circumstances dictated, very casually dressed, with no time for ceremony:

> The word had got around before the conference that a major general had been flown in to take over command and it is possible they expected a general in uniform with a staff entourage and the usual trappings of office. Instead, they found an individual in an old shirt and shorts who looked just as scruffy as themselves. Very soon we were all on first name terms – the situation was too serious to allow for any protocol or ceremony.

Stretton noted:

> Approximately 20 members of the Darwin Civil Defence and Emergency Service Organisation were on duty during the emergency. They were initially deployed to direct the public to various refugee centres and subsequently they assisted in the running of these centres.

Stretton's period of leadership was short but significant. In summary, he was gifted with the ability to make decisions fast, with a minimum of information to go on. In a lesser man, the chaos and lack of surrounding data would have caused the reverse: many another in the position would have simply frozen, without the courage to go forward. But the major general was to be the reverse of that.

Stretton was also blessed with another ability – that of keeping people informed. Within a day he had decided to make a daily radio broadcast, using the transmitters of the Australian Broadcasting Commission, as it was then, on the AM band. In those days the radio was just as prevalent in Australian society as it is now, but its presence was to be seen in a somewhat different way.

We now turn to an outline of what each of the three services did immediately following the cyclone, and for some time after.

Wrecked civil aircraft at Darwin airport following the cyclone, which was a shared RAAF facility. (South Australian Aviation Museum)

A Cessna 310 upended against a damaged Darwin airport hangar. Over two dozen civil aircraft were destroyed by the cyclone. (South Australian Aviation Museum)

CHAPTER 4

THE AIR FORCE STORY

At the time of Cyclone Tracy's turbulent passage, the Royal Australian Air Force in Darwin had long been under the control of Headquarters, Operational Command, responsible for all RAAF air operational activities and training in Australia.

Darwin's unusual airport status – which remains today – was that of an international airport operated on a joint-user basis by the RAAF and the Department of Transport. Some navigation aids and other facilities were provided by the Department of Transport but the provision of the bulk of the facilities was the responsibility of the RAAF.

The function of RAAF Darwin was to support deployments of operational units of the RAAF based elsewhere by providing communications, accommodation and other necessary facilities. It also functioned as an important staging base for RAAF aircraft and service aircraft of Allied nations.

Since WWII it had been used for the conduct of operational exercises by the RAAF either alone or in conjunction with air forces of Allied nations. It was frequently called upon to support rescue operations, mercy missions and flood relief throughout the northern areas of the continent and the seas to the north.

In 1988 RAAF Base Tindall was added to the Top End. Its location, 320 kilometres inland "down the Track" just south of Katherine, gave it "strategic depth" – in other words an enemy would have to get through ground-based defence systems and aerial fighters to reach it. The base was constructed in 1942 and was originally called Carson's Airfield. Nowadays, as it was in 1974, it hosts major exercises featuring visiting aircraft and personnel. The proximity to Delamere Air Weapons Range makes it a good location to conduct high-end training sorties for the RAAF and coalition partners.

At the time of Cyclone Tracy, the posted strength of RAAF personnel at Darwin totalled 670 of whom:

- 12 were in Headquarters, RAAF Darwin.

- 538 were in Base Squadron.

- 112 were in No. 2 Control and Reporting Unit.

- Eight were Service Police. Included in the total were three members of the RAAF Nursing Service, one WRAAF officer and 30 airwomen.

- Dependents totalled 913.

Before Tracy's onset, members of the RAAF had been briefed that as soon as the cyclone had passed, they were to attend to the immediate safety of their dependents and assist the injured. They were then to report at once for duty. Nevertheless, the acting RAAF commander Wing Commander WJ Monaghan reported the response to this order "exceeded expectations." This is all the more remarkable, seeing as over 200 of the 400 houses utilised by the RAAF were destroyed.

RAAF C-47 A65-104 which as described by Group Captain Hitchins "finished up in the front garden" of his house at the RAAF base after being tied down in a hangar. The aircraft was written off. (South Australian Aviation Museum)

As noted in Chapter 2, the actual RAAF commander, Group Captain D Hitchins, had been on leave 130 miles away on 24 December. Hitchins later recalled being picked up "late in the morning" at Smith Point by the C-47 that had initially been flown down to Tindall. He had listened to as much as he could of Darwin proceedings on a local ranger's radio, without being able to make much sense of it. He recalled:

> … the whole place was just like one vast rubbish dump. There were shattered buildings all over the place. I flew over our own house and had a good look at the loungeroom floor through the ceiling because it was all collapsed. It was just one great big mess of littered foliage, collapsed trees and components of buildings that had blown about all over the place like pieces of confetti. That's about all you can say. I don't think there was one building in Darwin that was undamaged.

Hitchins described the damage to aircraft:

> A civil DC-3 parked near one of them went straight across the top of the hangar. That would have to be 30 feet high, and the aeroplane went straight across the top of the hangar and finished up on its back … and one of our DC-3's, tied down in a hangar not far from my house, finished up in the front garden. A helicopter was flattened in that hangar when it collapsed. The small civil tarmac [had] 25/30 light to medium sized commercial aeroplanes were wrecked there, tangled up.

"Down south" at 0900 on Christmas Day the RAAF ordered a C-130 of No. 36 Squadron to be on standby, and 30 minutes later another, this time of No. 37 Squadron. A BAC 111 of No. 34 Squadron would carry Dr Rex Patterson, the Minister for the Northern Territory, to Darwin and – it was soon decided – Major General Stretton. This aircraft had left Fairbairn at 1530 also carrying "three surgeons, an anaesthetist, a registrar and three nursing sisters, as well as medical supplies."

A profile of C-130E A97-168, the first aircraft from the south to arrive in Darwin after the cyclone. (Juanita Franzi)

Initially flying to Mackay, the BAC 111 pilot learnt he was directed not to proceed beyond Mt Isa unless he knew the runways at Darwin were operational. The decision was made to merge the passengers and crew with one of the Hercules, which had also been loaded with medical personnel and supplies.

Keith Kershaw, the flight engineer on the Hercules, remembers their approach to the devastated northern capital:

> We began to feel the effects of the cyclone somewhere around Katherine as we began our let-down into Darwin. The method of approach was an ARA (Airborne Radar Approach) whereby the navigators used the aircraft radar to locate the runway. Because of the fact that the radar waves are bounced off the runway into the ether the runway appears on the radar screen as a big blank rectangle. We saw the 7 kerosene flares on the left-hand side of the runway at about 500 metres. The distance is a guess as rain was blurring vision and is based on the length of time to land from first observation of the flares.

> We landed firmly due to water on the runway, dead on centre line, and taxied to dispersal. Group Captain Hitchins met the aircraft and escorted the passengers and our captain to points unknown. The rest of the crew remained to configure the aircraft for the return flight and maintain communications with Air Force Sydney. The link was used by Group Captain Hitchins on more than one occasion. We had to refuel with engines running and were notified sometime before sunup, that the most seriously injured were to be loaded shortly.

In Darwin, at 2220 local time, the C-130E Hercules (A97-168), the first aircraft from the south to make it into Darwin, touched down in what were described as "extremely marginal conditions". Group Captain Hitchins was on hand to greet the first personnel into the city, although as he later recounted, in the civilian clothes he had been holidaying in and looking rather bedraggled by this time.

The aircraft's arrival also re-established communications south. Using the embarked radio, a call was put through to the RAAF's Air Vice Marshal Robey at his headquarters. The advice was that severe damage had been inflicted on a Dakota and the Iroquois; all of the hangars had

C-130E A97-181 which was the second Hercules to arrive in Darwin after Tracy as described by Flying Officer Jack Fanderlinden. It is seen at Richmond in 1977. (Nigel Daw)

An American C-141 Starlifter transport at Darwin.

been flattened, and 90% of the RAAF's buildings destroyed. Hitchins asked Robey for two helicopters and two Caribou transports.

The first night-time arrival of one of the giant Hercules aircraft was somewhat traumatic. Then Flying Officer Jack Fanderlinden recalls:

> My Flight Commander, Squadron Leader Bill Fewster, and I were the captains of the first two C-130E aircraft to launch out of Richmond on rescue missions to Darwin, when Tracy hit. Bill Fewster flew direct to Darwin … he flew the first Herc into Darwin after Tracy hit.

RAAF C-130 loadmaster Stuart Tarrier who flew into Darwin on 27 December 1974.

> I flew into Mt Isa before flying into Darwin. My co-pilot on that flight was Flying Officer Carl Sandford. I had to pick up rescue and relief gear (tents, etc) from Isa and fly it all to Darwin. With the stopover at Mt Isa, I arrived at Darwin at night. All the navigation aids at Darwin were unserviceable, there was no communication with anyone and there were no lights anywhere in Darwin (either in town or at the airfield) to help us identify where the airfield was.

The weather was not good, with low cloud and scud around. I used my radar to identify where the airfield was likely to be and made an approach. When we broke out of the low scud at around 200 feet on radar altimeter, we could see nothing, but we did see a red rotating beacon (one on top of a RAAF vehicle as it turned out) about 45 degrees out to our left. We made for the beacon with our aircraft landing lights and picked out the only cleared piece of the very long Darwin airport runway. The vehicle with the red rotating beacon was at the start of the cleared piece of the runway. There was about 4,000 feet

Exhausted RAAF C-130 crewmen resting between flights on the Darwin tarmac. (Ian Frame)

of runway cleared in the centre of the actual runway. Every other bit of the runway was covered in debris.

We were the second Herc into Darwin after Tracy hit. It was an exciting arrival – in fact, it was even more exciting than any of the arrivals I did in Vietnam during the war.

The RAAF medical team was moved to the Darwin Hospital where they started work with the superintendent and staff. A decision was made to evacuate the more seriously injured patients. Fifteen were selected, along with three from the RAAF Base hospital. This initial group included five paraplegics, one suspected case of gas gangrene, and others who had ruptured spleens or organs removed. They were flown out on a Hercules for Sydney at 0400 in pouring rain. Another selection was made, and these were also flown out; this time thirteen casualties and 85 civilian evacuees. This Hercules took off for Brisbane. It was just the start of a continuous stream of both service and civilian aircraft.

The response from other countries was swift, and in most cases, it was initially in the shape of aircraft. One grateful resident, Ludij Peden, noted:

America … is always the first to help. They sent in their giant Starlifters, a plane that was umpteen times larger than anything Australia possessed. Indonesia also sent as many planes as needed. The rest of the airlift was undertaken by Hercules, Air Force transports, commercial airliners and small aircraft constantly landing and taking off, bound for all major airports around the country. Because the airport is almost in the centre of Darwin we watched fascinated as these big American monsters took to the skies.

Stuart Tarrier, then an RAAF sergeant with No. 36 Squadron, and working as a loadmaster, recalls one evacuation. His aircraft flew into Darwin on 27 December, initially loading two generators onto their Hercules in Townsville for the trip north. Stuart noticed "bits of corrugated iron and so on around 50 miles from the runway."

Their return flight south was an evacuation with 143 passengers, who were mostly women and children. Stuart remembers:

They had been at the airport for hours – there were no facilities left there. We loaded up and put some babies two to a seat. We took off with just me down the back trying to heat up water for babies' bottles.

The Hercules pilot soon received a message from the RAAF about an offer from Mt Isa, probably from the Salvation Army. Stuart recalls it was:

… to give us medical assistance for all patients who needed it, and food, clothing and so on. It took a long time to get everyone off, and I was last to only get a cigarette and a cup of coffee. Then we re-boarded and it was another four hours to Sydney.

Jim Naylor, a loadmaster on another Hercules transport, recalled babies being a concern for him in a number of ways. On one trip south, apart from a pregnant woman thinking she was not going to make it to the hospital in time, the last passenger he helped aboard was a new mother with her baby wrapped in towels in a washing basket. On arrival in Adelaide a journalist tried to get on board the plane to take photos of the baby and Jim pushed him back down the loading ramp "with the imprint of my boot." He was concerned he would get into trouble for that but heard no more about it.

P-3B Orion A9-292 at Edinburgh, South Australia, in April 1975. On 27 December 1974 this aircraft searched for missing fishing vessels in waters off Darwin. After landing to refuel it embarked a dozen evacuees for the return flight south. (Nigel Daw)

On another flight his Hercules was radioed by a US transport also taking passengers south. They had just had a birth on board and wanted to know what the "vertical international boundary" was, to determine whether the new arrival was a US citizen or an Australian.

RAAF officer Ian Frame remembers the call to get going:

> On Christmas Day 1974, I was a junior captain at 37 Squadron and was settling down on my first ever Base Duty Office duty (being an unmarried 20-something boggie I was fair game for the squadron programmers).

> At exactly 10am, Flying Officer Randall Kingsley, the Base Orderly Officer and an Air Transportable Telecommunications Unit (ATTU) engineer, and I were settling in front of the TV in the Sergeants' Mess (a combined mess due to Christmas) watching the Pope knock on St Peters Cathedral door (hence I can remember the exact time), when Randall got a phone call. He came back very perplexed, stating "I have been asked to get all our deployable gear ready, but they couldn't give a reason except something about Darwin".

> Thirty minutes later we were up to our ears in directives, phone calls and, soon afterwards, reporters (all of whom appeared to know the Chief of the Air Staff, apparently, but refused our offer to get him on the phone.)

> Chaotic day, highlighted by how many personnel, including e.g. air movements personnel, switch operators, chefs and 486 Squadron maintenance personnel, called in or just turned up ahead of recall. 37 Squadron even had two loadies picked up from their fishing camp on the Darling River by police boat, much to their surprise.

> I went flying to Darwin first up the next day, Boxing Day, as the third C-130, taking Randall and his ATTU component. My first flight from Darwin was evacuations to Brisbane, then up

Evacuees embarking on a Boeing 727 airliner. Such aircraft soon joined the initial effort by RAAF transports. (NT Library)

to Townsville to pick up electrical linesmen and their vehicle. As the first "Darwin" aircraft to reach Townsville I was besieged by reporters. I flew A97-178 then, later, A97-190, with evacuees and the occasional pet, mainly to Brisbane and Sydney, with return to Darwin flights diverting as necessary e.g. to Albury to pick up generators. Very, very long crew days.

The first days however had not gone so well. Hitchins had an argument with Stretton, as the normal RAAF chain of command had not been superseded:

I had no intention of standing around putting up tents for people who were well capable of putting up tents for themselves. He wasn't very pleased about that. His face became red, and he said something about not wishing to have any confrontation with me and I think I said, "Yes, I don't want any confrontation with you, but I'm sorry I can't accept orders from you."

Stretton later wrote:

At the RAAF Base … my directions about the preparation of a transit area seemed to be slow in being implemented and there still seemed a reluctance by the RAAF to give any assistance outside the area of the RAAF boundaries. The OC RAAF obviously still had some doubts about my authority so later that day I contacted Canberra and had the matter resolved.

Later Hitchins said:

I was a little disappointed when he came around giving me orders. It wasn't the first time I'd had Army officers in other places attempting to give orders to Air Force people. It is not unheard of. It has been tried before, and I saw this as just another attempt for the Army to spread their authority over all that they could survey, and my reaction was the same as it's always been. I have no particular objection to the Army but I don't take orders from them. And what a fool I'd be. I mean, where would you be if you started taking orders from every twerp that walked into the gate. You'd finish up in chaos and you'd never do anything.

In southern centres, the RAAF response gathered speed, cancelling the usual Christmas break. It was focused in the main around Nos. 36 and No 37 Squadrons, operating C-130A and C-130E Hercules transports. Sometimes this led to unusual incidents. An aircraft navigator:

… recalled to duty from Pioneer, Tasmania, set out by car to Launceston airport. In an effort to catch a civilian aircraft he travelled fast in his car and was stopped by a police patrol. When he explained the situation, an understanding police officer provided him with an escort to the airport.

Following a request on 28 December, an RAAF Canberra flew a mapping survey mission, with the Army Survey Regiment providing the map sheets. A four-engine P-3B Orion maritime search and strike aircraft carried out a search for four missing fishing vessels on the afternoon of 27 December. The flights backwards and forwards gathered strength. Dave Jones, who was an acting-sergeant loadmaster with No. 36 Squadron at the time, recounted:

I was living in Richmond on Christmas Day when my flight sergeant knocked on the door and asked if I had been drinking. As I was on leave, I was quite rude to him! We were to fly to Darwin the next morning. From memory we took such supplies as disposable nappies with us. On the return trip we carried about 110 passengers south.

On the morning of the 30th we flew down to Laverton to pick up some diesel generators

C-130A A97-216 seen in April 1975. The fleet of C-130A and C-130E Hercules were the backbone of the RAAF Darwin relief operations, contributing almost 1,200 flying hours. (Nigel Daw)

and some pallets with 144 small generators on them. Dave Hitchins, who had once commanded my squadron, told us to take the aircraft down the back of the airfield and to get some sleep. We then did a medical evacuation load up procedure. At midnight mid-flight one of the doctors got promoted and we toasted that in orange juice.

The services the ADF could provide were indeed varied. Photographer Ken Markwell stayed on until February photographing everything for the ADF. He took photos of damage around the base, and then in helicopters over Darwin. He was using a Single Lens Reflex camera, and developing some film himself with the chemical process of the time when he could, but most film was sent to RAAF Amberley on the shuttle runs.

The number of decisions on an enormous variety of matters, and the speed at which the decisions were made are stunning. For example, in one Operations Log alone – for 28 December – a single word "yes" to 22 tons of oranges to be transported from Adelaide to Darwin was made; a rumour of typhoid being published in the *Age* newspaper was negated; and that NOCNA be made the temporary harbourmaster was approved.

The power of armed forces logistics was often in evidence: "General has directed that 100 tons [sic] of roofing iron at Amberley to be flown to Darwin tomorrow." Some items suggested for ADF transportation though were rejected: "WD & HO Wills have made an offer … of 250,000 cigarettes." They were advised they were not required.

On 26 December, a replacement Iroquois helicopter was transported north by C-130. Another flew in 7,000 blankets, 6,000 pounds of milk and 2,000 pounds of Red Cross medical supplies. While the run of supplies north was enormous, the evacuations were equally as large. On Boxing Day, the RAAF flew out 680 persons, and the following day 1,218 in 22 sorties. The Australian aircraft were joined by a USAF C-41 and an RNZAF Hercules that also carried evacuees. The Indonesian Air Force flew in supplies by transport aircraft, and the Royal Air Force from Britain utilised aircraft based in Singapore.

Although the aircraft of the armed forces flew ceaselessly, the enormous capacity of civilian airliners quickly supplemented them. One of the giant Boeing 747 machines, with its distinctive second deck up and behind the cockpit area, managed to fit 674 people on board on a flight to Sydney. Interviewed years later, the pilot Don Howe said "It was a case of get on and hold on … every adult had a child on their lap." Forty years later, a plaque was unveiled to the record-breaking civilian evacuation flight at Darwin Airport.

Not everything about flying out on a military plane was positive. Jan Anderson recalled of her flight south in an Australian Hercules on New Years Eve:

> We were only allowed one case. It was an awful flight. I was sat right at the back where they had all the luggage strapped into nets. It was cold in Adelaide, but then again I only had on shorts and a T-shirt.

Mick Taulelei remembers his flight south chiefly because he had to sit sideways on the Hercules' webbing seats. He was aged 14 and travelling with his elder sister Angie and his two younger brothers Phillip and George. They flew to Adelaide, where he recalls on landing "They gave my sister some money, and then we were taken to a centre where we could choose clothes from big heaps." Mick managed to obtain the list where the four of them, flying under their mother's maiden name, were ticked with a pen mark.

Ludij Peden described her evacuation flight which she says turned into a torment:

> An hour into their flight the Hercules was struck by lightning. The pilot assured the passengers not to worry but that they had to turn back to Darwin. All the instruments and communication systems had been blown out by the lightning. The plane droned on another seven hours and then landed.

> As they disembarked, the passengers realised that they were not in some distant other city but back in Darwin. They had assumed that since they'd only been an hour out, when the lightning struck, and that they had flown another seven hours, that the captain had changed his mind, and decided to fly on.

> Again, it's these sorts of events that bring out the best and the worst. When it dawned on the passengers what had happened, the hysterics started. After all that had occurred over the last few days, tolerance was low. Women were screaming at the crew that they all could have been killed; the crew couldn't be trusted; and that they wouldn't fly with them ever again. Our little boys were wide-eyed and concerned amidst the hysteria. Grandma kept her cool and went up to the pilot. She thanked him for safely bringing them down. She would happily fly with him again for he obviously was very skilled.

Hitchins described taking the initiative with spraying Darwin from the air to prevent disease:

> … the hygiene problem was extensive. Anyway, Sergeant Cowan went around and had a look, and he came back and told me that if we didn't do something fairly soon to prevent the menace of breeding flies and so on, we were going to be living in the middle of a very nasty situation.

> I couldn't contact Charles Gurd … so a little bit of sculling around, and found three old gentlemen, mostly old fellows about my age, who were agricultural pilots … we gave those blokes cart blanche to spray Darwin, including the RAAF Base. We said we'd provide them with a place to live, free beer, and they could control their own air operations.

> Without pay, the three then "spent the next three or four days beating up and down the main streets of Darwin at about 100 feet".

The pilots eventually did get paid, said Hitchins. RAAF operations continued in all sorts of ways. The Navy helicopters, recalled Hitchins, worked together with Air Force aircraft in a co-ordinated system. He also refers to "about 3,000 people" of the Air Force who were posted to Darwin for "a three-month period and then returned to their normal bases in other parts of the country."

The aircraft hours flown by the Air Force in the relief operations were dominated by the four-engine Hercules transports. What the Macchi – a lead-in jet fighter – did is unknown, but almost a dozen different types contributed the flying hours detailed in this table on the right:

Aircraft hours flown by the RAAF in Tracy relief operations	
Hercules	1,197.8
Caribou	102.5
HS748	135.0
BAC 111	61.7
Neptune	5.9
Orion	31.8
Canberra	15.9
Iroquois	6.3
Mystere	43.7
Macchi	3.2
Dakota	29.3

One account from the State Library of South Australia has it that 25,628 people were evacuated by air, and 7,234 left by road. By 31 December 1974 only 10,638 people remained in Darwin. However, other accounts concerning the numbers evacuated vary. Especially in the first week after the cyclone, it was more what was done rather than a record being kept of what was done that was important. People could also simply leave in a car of their own accord. The number of people living in Darwin before Tracy was not precise. Therefore, the exact evacuation and remaining resident numbers are likely to remain amorphous.

Now we turn to the Navy's story.

Evacuees crowded inside an RAAF C-130 Hercules.

CHAPTER 5
THE NAVY STORY

The Navy operated two major bases in Darwin, but they had an unexpected degree of complexity. The patrol boats were based "in town" at the wharves. The Navy's administrative centre was HMAS *Melville* on the clifftops above. Its communications base of HMAS *Coonawarra* was fourteen kilometres away down the Stuart Highway. New facilities for communications aspects were under construction in Shoal Bay.[1]

The naval bases were manned by 305 naval personnel, including 90 WRANS, plus an additional 29 civilians. Sailors afloat numbered 75 personnel. These 409 people had 287 dependents, giving the Navy a presence of 696 in a city the population of which was 46,656 on 30 January 1974. During the following month, the Royal Australian Navy would embark upon its largest peacetime disaster relief operation, involving 13 ships, 11 aircraft and some 3,000 personnel.

Immediate response

As we have seen, *Arrow* was lost. Bob Dagworthy's command was sunk, two men were dead and his life and those of his surviving crew would never be the same again. More will be seen of Bob and the *Arrow*.

The front gates of HMAS Coonawarra on the Stuart Highway in the days after the cyclone. (RAN)

---

1   In 1984 a new maritime base with a basin and a syncrolift facility was opened actually inside the Army base of Larrakeyah Barracks in town where they remain today.

RAN HS748 N15-710 which flew a naval dive team into Darwin on 26 December 1974. (Nigel Daw)

The naval personnel then based in Darwin possessed only a limited capability to render immediate assistance to the stricken city and its community. Of the four Darwin-based Attack class patrol boats of the RAN, *Arrow* had sunk under Stokes Hill Wharf with the loss of two lives, *Attack* was driven ashore at Doctor's Gully by the sheer force of the cyclonic winds, and *Advance* and *Assail* were both damaged. Darwin Naval Headquarters was destroyed, as was 80% of the patrol boat base and 90% of the naval married quarters. The oil fuel installation and the naval communications station HMAS *Coonawarra* were extensively damaged.

Initial relief was limited to search and rescue operations on the harbour foreshore and in waters out to Melville Island. Communications facilities in Darwin, both military and civil, were crippled, and initial communications were dependent upon Army mobile terminals and the communications systems in *Advance*, *Assail* and the motor vessel *Nyanda*.

The Navy – the voyage north

Although it was Christmas Day, naval personnel in the southern cities and towns quickly mobilised themselves or responded to telephone calls. Lieutenant Commander John Simmons recalled:

> From Sydney I caught an RAAF Hercules from Richmond to Darwin via Adelaide and sat on packing cases for most of the way. The ships had just arrived and the relief effort was in its very early days. I joined *Stalwart* and became fully involved in providing dry accommodation for Darwin residents. I was overseeing roofing and maintenance works but was mainly involved in the identification of worthy projects for the naval teams to undertake. One memorable project was the roof repairs to the Parap Hotel where the repair teams worked in two shifts. One on the roof in the blinding reflections of the sun from the galvanised iron and the high temperatures, whilst the other retreated to the drive-in bottle shop cold room to cool down and replace their body fluids.

As the gravity of the disaster became apparent, a naval task force, under the command of the Flag Officer Commanding the Australian Fleet (FOCAF), Rear Admiral DC Wells,

CBE, RAN, was assembled to render aid to Darwin. A general recall was issued to all personnel. Approximately 50% of all Sydney-based ships' companies were on annual leave, with many interstate. Of the 2,700 personnel on leave, 2,200 were able to return to their ships prior to sailing, and others subsequently managed to join their ships in Townsville. Volunteers from other Sydney-based ships and establishments filled the positions of those who could not return to their ships in time. All manner of stores were embarked on the deploying ships, ranging from combat bridges, vehicles and building materials down to disposable cutlery. Commodore Guy Griffiths, captain of the aircraft carrier *Melbourne*, recalled later that:

> … nobody complained about Christmas leave being interrupted. It was just part of Navy life; you're called to assist people; and you do it to the best of your ability.

The carrier HMAS Melbourne approaching Darwin in January 1975, with six Wessex helicopters on deck.

The response of Operation *Navy Help Darwin* was swift. The first RAN asset to arrive in the disaster-stricken city on 26 December was an HS748 aircraft from No. 851 Squadron, carrying blood transfusion equipment and a team of Red Cross workers. A second HS748 aircraft carrying members of Clearance Diving Team One (CDT1) arrived shortly thereafter.

On 26 December:

- Landing Craft (Heavy) HMAS *Balikpapan* and HMAS *Betano* sailed from Brisbane.

- Survey vessel HMAS *Flinders* sailed from Cairns.

- Aircraft carrier HMAS *Melbourne*, destroyer HMAS *Brisbane* and destroyer escort HMAS *Stuart* sailed from Sydney.

Landing craft were – and are – very useful in disaster situations. They basically are a big flat-bottomed ship with a bow ramp which may be lowered onto a beach. In situations where the local wharves have been damaged or have wreckage making them unusable landing craft may be simply driven onto the nearest beach. They are big enough to carry bulldozers to make a dirt road from the beach if necessary, and thereafter heavy trucks to take materials back and forth.

Four S-2E Tracker aircraft from Nos. 816 and 851 Squadrons prepared to fly to Darwin but were placed on standby and eventually stood down. The following day:

- HMAS *Hobart*, HMAS *Stalwart*, HMAS *Supply* and HMAS *Vendetta* sailed from Sydney.

- HMAS *Brunei* and HMAS *Tarakan* sailed from Brisbane.

Nine Wessex helicopters from Nos. 817 and 725 Squadrons were embarked in *Melbourne* and *Stalwart*. HMAS *Wewak* subsequently sailed from Brisbane on 2 January 1975.

The Executive Officer of the carrier *Melbourne*, later Commodore Thomas Dadswell, AM, RAN (Rtd) – abbreviated further to his ubiquitous nickname of Toz – later reflected:

> I was on leave at the time having just completed three years on the Defence Joint Staff. On the morning of 25 December, the decision had been taken to sail the Fleet to Darwin to assist in the clean-up and to render what aid we could. As my new posting was as XO *Melbourne* I rang DCNS and asked if I should join the ship. His reply was to the effect that no captain needed two executive officers so I should enjoy my Christmas dinner. Within the hour another admiral rang, and directed me to get myself to Richmond air force base as soon as possible as I was required in Darwin. That was the full extent of my briefing. I left Richmond on a C-130 the next day arriving in Darwin later that day.

> For the first few days I was involved with the evacuation of civilians but then a message from the *Melbourne* appointed me as Officer in Charge, Shore Headquarters Darwin. Having given me a title, the next message requested me to forward my proposed plan for the clean-up. As the book of instructions on how to clean up a city had been blown away in the storm I had to start with a clean sheet. It was an interesting exercise.

Navy News later reported:

> Every possible contingency was thought about, and in some cases acted on. For example, some of the ships built catamarans, in case they would be needed for offloading people or supplies. They also built and painted street signs and set up a general overall organisation for getting onto shore.

> On board HMAS *Melbourne*, the men set about the giant task of grouping all the stores for Darwin into 2,000-pound weight [907 kilogram] bundles so that the helicopters would just hook onto a bundle and be away. Each bundle was put ready on the flight deck, and in the hangar.

> A complete Disaster Co-Ordination Centre was set up in the Briefing Room onboard the flagship. This was where everyone going ashore from *Melbourne*, and what they were doing, was controlled.

> Every day, as the ships steamed up the Queensland coast, there was a close re-examination of special needs of Darwin. If it was found that there were any requirements for stores which were not onboard, signals were sent back to Sydney and the stores were flown to various pick-up points on the coast. As the ships passed, the stores were flown to the ships.

The ships arrive

As we have seen, the Director General of the National Disasters Organisation, Major General AB Stretton, DSO, had arrived in Darwin on 26 December with his staff officer to establish an Emergency Services Organisation Committee. Captain Eric Johnston, RAN, Naval Officer Commanding the North Australia Area (NOCNA), was appointed to the committee as Port Controller, with responsibility for controlling the port and its approaches, and for drafting an Emergency Plan in the event of a further cyclone.

As preparations were made for the arrival of the naval task group, Captain Johnston relocated the naval headquarters to his residence, Admiralty House. Following an

Captain Eric Johnston was the senior naval officer in Darwin during Cyclone Tracy. In the 1980s he became Administrator of the Northern Territory.

exchange of signal traffic between FOCAF and NOCNA, it was agreed that the RAN relief force would be allocated responsibility for clearing and restoring 4,740 houses in the northern suburbs of Nightcliff, Rapid Creek and Casuarina. HS748 aircraft continued to ferry personnel and stores to Darwin and evacuees south. Evacuees were accommodated in HMAS *Kuttabul*, HMAS *Penguin* and HMAS *Watson* in Sydney; and HMAS *Moreton* in Brisbane. Clearance Diving Team One was surveying damage to the patrol boats and civilian craft, searching for missing vessels, clearing Stokes and Fort Hill Wharves, and assessing how to extract the wreck of *Arrow*.

The first ships, *Flinders* and *Brisbane*, arrived in Darwin on 31 December. *Flinders* surveyed the approaches to Darwin to ensure the safe passage and anchorage of the Task Group, while *Brisbane* landed working parties and established communications with NOCNA. *Melbourne* and *Stuart* arrived on 1 January; *Stalwart* on 2 January; *Hobart*, *Supply* and *Vendetta* on 3 January; and the landing craft (heavy) *Balikpapan* and *Betano* on 4 January. Landing craft (heavy) *Brunei*, *Tarakan* and *Wewak* arrived the following week on 13 January. The ships had brought with them some 3,000 naval personnel.

At one stage the inclusion of a British submarine, HMS *Odin*, which was attached to the Australian Submarine Squadron pending the delivery of the RAN's final Oberon-class submarines, was envisaged to connect to the city's electrical power supply. However, after some investigation it was found the necessary connection would be too difficult. Another problem developed on 5 January: the freighter *Lake Illawarra* collided with the Tasman Bridge in Hobart, killing 12 people as a central section of the bridge collapsed on top of the ship, sinking it and taking cars and people with it. This meant a further deployment from the forces, with a dive team and landing craft the first demands.

The arrival of *Melbourne* precipitated the establishment of a Shore Command Headquarters (SCHQ) at Admiralty House to coordinate the working parties, which were tasked by the Emergency Services Organisation. Working parties were typically composed of ten or fifteen officers and sailors, depending upon the nature of the task.

A profile of Wessex N7-200, one of nine that were extensively utilised during Operation Navy Help Darwin. (Juanita Franzi)

The *Melbourne* took seven Wessex helicopters to the north with her. Later another two arrived on HMAS *Stalwart*. The helicopters "worked as aerial cranes, as people-ferries, as freight vehicles, as search and rescue craft and as reconnaissance helicopters." They were quick movers, with take-offs from the flight deck even as the carrier was coming past East Arm into the harbour. A small fleet of boats was also lowered to the sea: "Within 60 minutes, more than 400 men were ferried ashore from the carrier."

Darwin resident Barbara Conje recalled:

> In the morning after the cyclone a group of sailors turned up at the wreckage of the house volunteering to help. By then we'd done what little we could, but their arrival was a bright spot in the morning and much appreciated.

Military historian Paul Rosenzweig later wrote:

> Eric Johnston co-ordinated the recovery effort from the sixth floor of what became known as *HMAS MLC*, the Mutual Life and Citizen's Assurance (MLC) building in Smith Street.

> He took up his post as naval officer commanding north Australia … and was heading for an end to an uneventful term in Darwin when he was literally caught in the rubble of naval headquarters, destroyed by Cyclone Tracy on Christmas Eve 1974. Johnston and two sailors crawled out.

> Over the next few months, he played a leading role in handling the ensuing emergency and clean-up of Darwin. The Navy under his direction performed an enormously successful job in the demoralised, depopulated city. Johnston was awarded membership in the Military Division of the Order of Australia in 1975 for "… outstanding leadership, exemplary conduct and steadfast performance of his duties while exposed to the dangers of Cyclone Tracy and for his dedication and tireless efforts towards and for his restoration of Darwin's defence and town services."

Johnston was no flash in the pan leader either – he had a history of strong service behind him. One comment on his previous service enthused:

> … a very fair and passionate naval officer, whom the crew loved. He ensured that the ship was at all times ready for operations … a true sailor, he kept the ship humming. Sailors would follow him whatever the state of the ship or sea.

Stretton, in his broadcast to the people, said that the arrival of the Navy was "the biggest convoy of Australian vessels that has ever put to sea together since World War II" and furthermore

A Wessex helicopter at work in the wreckage of Tracy - note the man underneath on the winch. (RAN)

that "… the Navy commander constantly acted on his own initiative without the necessity for any orders to be given."

Johnston noted of his own people:

> Despite this, the cheerfulness and the work of the sailors and officers was a subject of much favourable comment. I mention officers for, apart from those controlling personnel, the officers were best employed in manual labour and a photo of the captain of *Melbourne*, Commodore Guy Griffiths, stripped to the waist and carrying a large baulk of timber, made the front pages in southern newspapers.

The naval helicopters

The helicopters on board *Melbourne* were especially useful. They could be used as a truck, lowering anything – people or pallets of supplies – to wherever they were needed.

> The authorities were able to get out beyond the city limits to look for missing or stranded persons and searches were conducted all along the coast searching for missing trawlers. Once the searching was completed the helicopters settled down to a role of providing the capability only a helo can. They were used to move people quickly from place to place. They were able to transit across areas that were blocked to road movements. They could get people into places that were otherwise inaccessible.

Pilot Geoff Ledger recalled he:

> … flew the first military helicopter into Darwin off HMAS *Melbourne*. We took off at first light around 50 miles east and flew in to pick up Captain Johnston who was NOCNA at the time. I can still recall the absolute devastation of the entire area. We were frantically looking at our map to endeavour to locate reference points for our flight to NOCNA's house. The sight that morning that I found difficult to comprehend, was how little pockets of houses and vegetation actually escaped the jaws of Tracy. Everywhere else was flattened like the photos I recall of Hiroshima.

A Wessex lowering a slung load, possibly air-conditioner machinery, onto the roof of a hotel.

> I stayed in Darwin for six weeks. HMAS *Melbourne* was anchored in the harbour and unloading many tons of equipment and stores. I was one of the Wessex helicopter pilots from HS 817 Squadron which flew many hours in support of Operation *Navy Help Darwin*. Life onboard *Melbourne* was very hectic, it was like a small city in full swing 24 hours a day. Work parties would depart at first light in LCHs, and the helicopters would begin some tricky external load operations early before the breeze got up. Most of the squadron crews flew for two days, worked ashore for a day and then had a day off for personnel needs and administration. These so-called days off never really eventuated because there was always additional work to be done in the hangars.

> The flying conducted and work party tasks were very rewarding and hard, hot work. I can recall one sortie, when we successfully delivered a load of corrugated iron sheeting to the top of the Territorian Hotel. Just after releasing the load, our main rotor struck an exhaust stack on the roof, and I had to gingerly fly the helicopter away from the roof to a nearby park to repair two badly damaged rotor blades. Operational hazard!

> After three weeks, HMAS *Melbourne* had to depart to continue operational work ups. I was transferred to HMAS *Stalwart* which stayed to continue restoring services to Darwin. Our flying rate was maintained, and we were now involved in flying out to some of the outstations and island communities. This was a most satisfying experience, some of these people had been out of contact with the real word for many weeks.

Sometimes the use of helicopters was rather unusual. A local hotel was housing people for the emergency, but the windows as was normal in many a hotel were not designed to open, and the humidity was around 90% in temperatures over 30:

> The air-conditioning plant on the roof of the Travelodge Hotel was damaged and there

were no cranes available to lift the replacement machinery onto the roof. It was expected that repairs would take weeks to complete. But a helo lift solved the problem much to the relief of the many people housed in the building.

Dave Jones, then a naval aircrew member, recalled that accidents were possible, and sometimes only narrowly averted:

> The Wessex was a great helicopter and very capable but it was single engined and engine failures for one reason or another were fairly common in the 1970s. An engine failure at height allowed the pilot to glide (autorotate) to a safe landing in a clear area if they judged it well. Engine failure at low level would be disastrous.

> Most of the flying at Darwin was low level, especially the load-lifting and the transport of work parties across the harbour. On one day a Wessex was hovering over the deck after dropping off a team of sailors and the engine failed, it landed hard on the deck with no damage. A few seconds either way would have resulted in serious injury or even loss of life.

Arrow and the Clearance Divers

One of the main problems that faced Darwin was the sinking of vessels near the main wharves. This was where most of the heavy equipment – bridge girders, heavy vehicles, mass pallets of materials and so on – would arrive. Therefore, such sunken shipping was a danger: the hulls and masts could snag or penetrate hulls; cables and ropes could entangle propellers, and leaking fuel could contaminate – just to name a few possible hazards. Ironically and rather sadly the Navy's own incoming expert dive teams found themselves working on one of their own service's ships, the sunken *Arrow*.

Duct tape is used to identify a vehicle, which has lost its windscreen, utilised by the navy divers.

With the arrival of the Task Group therefore, the primary focus for Clearance Dive Team One turned to the extraction of *Arrow* from near Stokes Hill Wharf, a task achieved on 13 January after much work. Captain Johnston went down to see what was happening, and encountered a journalist being a nuisance:

> The diving team had worked tirelessly. They had undertaken general harbour search and surveys, they had cleared the wharves and immediate approach areas and they had salvaged and raised the wreck of HMAS *Arrow*. I believe that the only time I lost my temper during the whole clear-up operation was when a well-known television commentator, encountering the clearance divers during one of their rest periods, asked them why they were bludging. This man does not know how close he came to joining the other debris in the harbour.

Writing later in his autobiography *Bubbles, Booze, Bombs and Bastards, A Clearance Diver's Story*, diver Larry Digney said Captain Johnston shouldn't have called the TV personality "a man" as he was a "low life weasel". All up, Digney said, the divers worked for 23 days

The hulk of HMAS Arrow after it was raised by Clearance Dive Team One.

straight, 12 hours on and 12 hours off. Diving, as the author can attest to from his own experiences as a divemaster, is an extremely strenuous sport, but non-stop work in low visibility conditions is much more psychologically demanding and physically draining. Only the extremely fit, brave, and capable Navy divers could have carried out the tasks they did.

The Navy's Official History of *Arrow* later concluded:

A Wessex amid suburban wreckage.

> Clearance Diving Team One successfully refloated *Arrow* on 13 January 1975 by attaching pontoons to her hull and using tugs to pull her clear of the wharf at high tide. She was towed underwater to shallow water at Frances Bay where she was surveyed, written off and later sold to a local businessman whose intention was to rebuild her as a museum piece.[2] The restoration process proved too expensive, and *Arrow* was eventually broken up where she lay.

Again, rather sadly, the pieces of the ship were buried as landfill. Nearby a shipping yard for the maritime contractor Perkins was being constructed out from the shoreline. Fill for the yard was needed, and that was where the pieces from the patrol boat were buried.

Deciding on the Task ahead

Navy News later reported on how a program for the naval shore parties was organised on the journey north:

> Commander TA Dadswell was put in charge of this headquarters. He was responsible for the overall operation of the headquarters, for liaison with NOCNA regarding the work,

2   The businessman was Stan Kennon.

and the direction of ships and special working parties.

The organisation of the shore headquarters was roughly divided into the following: Planning, Operational, Communications, Transport, Stores, Public Relations, Boat and Air Traffic Control. From the start it was made quite clear that the only way to operate was through direct requests from the public. So the Navy only took on jobs which were specifically asked for. The requests were channelled to Operations, where priorities were decided.

A work party of sailors burying spoiled foodstuffs.

The planners then fitted each request into the programme to be done as soon as possible. Every morning, Captain Johnston and Commander Dadswell went to the daily Town Disaster Central Committee at the Police Station. It was here that policy decisions were taken.

The Navy asked for, and was given, three suburbs to clean up, Nightcliff, Rapid Creek and Casuarina, which were the worst hit areas of the city. Commander Dadswell, shore-based controller of the relief operation, explains: "The men were divided into teams of about fifteen each. Each team searched the rubble and cleared up generally. Any valuables are marked and sent to HMAS *Stalwart* for safe keeping." This operation alone was a mammoth one like all the naval operations in Operation *Navy Help Darwin*.

What was actually a valuable item was "usually in the category of personal possessions such as legal papers and photographs, money, precious metal and the like." Here they were sorted, cleaned, marked, registered in a big book, sealed in envelopes or bags, and taken by truck to *Stalwart*.

The Supply Officer of *Stalwart*, Commander G Heys, says: "We hope that the goods can be turned over to a civilian committee as soon as possible for return to the rightful owners. Obviously, this is a big task and it is going to take a long time."

Furniture too big to move was to be stacked neatly in the centre of the property ... any building materials that could be used were cut into shape by Naval Shipwrights from the Fleet and stacked in piles for re-use. Commander Dadswell said: "There was a desperate need for fibro sheeting, so any of this found was stacked also for re-use."

Not everything was straightforward. Lieutenant Commander Simmons recalls:

There is only one bad experience which comes to mind and that involved an [Ed: deleted] who attempted to get me to replace the roof on her house (there was nothing wrong with the roof and there had only been a few minor leaks through the cyclone) with threats and coercion. Johno Johnson, with whom I was working, stood by me on that one.

The organisation and plans for the clearance of individual home sites was to remain the pattern for the ensuing three months and was adopted by the Army after these clearing tasks were handed over by the Navy. Darwin's mayor, Harry "Tiger" Brennan, later expressed his appreciation saying "We owe the Navy the greatest debt of all," a sentence that became a front page headline in the 17 January issue of *Navy News*.

The Navy's most unpleasant task

Cleaning up Darwin could hardly be said to be easy. It was the worst time of year for the weather – right in the middle of the Wet season, with daily high humidity only occasionally broken by torrential rain. The nature of the necessary tasks was also mostly unpleasant. John Riddle recalled:

> I always felt sorry for the Navy guys who got the job of cleaning out places like Woolworths Food Stores, and those sort of things. They were rotten jobs to do. They did a lot of that sort of thing, and then did a lot of house clearing and that sort of thing. Then the Army moved in and took over a lot of that work. But the Navy boys certainly copped some dirty jobs in the early time.

Navy News later reported:

> The Navy was asked to take on the revolting job of getting rid of all the rotting foodstuffs in freezers, refrigerators and cold stores. Food doesn't last very long in hot humid Darwin at the best of times. So it was of the greatest urgency to clear the rotten food immediately.

A work team from HMAS Vendetta, suitably attired for the Darwin climate.

Working parties started on this task immediately. It took them about a week. Because the task was so unpleasant, the parties were constantly rotated.

Dr Jonathon Hancock, in charge of the public health situation, later commented:

> The Navy, I know, cleared the Casuarina cold store, and had to use breathing equipment to get into it. You remember, this was Christmas Eve, and every fridge was full of turkeys, and hams, and all of the Christmas trappings which, in that humid wet season weather, just putrefied at a rate of knots.

Bob Luxton, then a naval lieutenant, recalled "the assumption seemed to be that we would come across numerous decomposing bodies while we were working. Fortunately, that didn't happen."

Dawn Lawrie, later a member of parliament, said:

> The Navy will forever remain heroes. We heard the Navy was coming within a matter of, say 24 hours, and it sounded too good to be true. But the Navy did come, and their immediate task was to start cleaning up, cleaning up the food which was lying – I mean fridges and freezers were strewn all over Darwin. And cleaning debris from people's yards and stacking it so that we could get into our own premises and assist in the clean-up.

> The Navy guys worked – this is an exaggeration, but say 23 hours a day. They worked until they dropped, and again, they were so kind and supportive of the few locals that were left that it was just wonderful. And I remember saying to one of them: "We won't get over this" – it's bringing it all back – and he said: "We're your Navy."

The Heavy Lift Capability of the Navy

The raw statistics amply illustrate the magnitude of the relief work undertaken by the RAN. Between 1 and 30 January naval personnel spent 17,979 days ashore, with up to 1,200 involved at the peak of the operation. Working parties cleared some 1,593 blocks and cleaned up schools, government and commercial buildings and recreational facilities. They installed generators, rewired houses, repaired electrical and air-conditioning systems, re-roofed or weatherproofed buildings, and maintained and repaired vehicles. Some parties worked to save rare plants in the Botanical Gardens.

Hygiene parties disposed of spoiled foodstuffs from houses, supermarkets and warehouses. Female personnel from *Coonawarra* supported civil relief organisations and manned communication centres. One enterprising sailor from *Hobart* filled in as a relief disc jockey for the local commercial radio station. The Wessex helicopters transported 7,832 passengers, 244,518 pounds (110,912 kilograms) of freight and made 2,505 landings. The HS748 aircraft completed fourteen return flights to Darwin and carried 485 passengers and 50,000 pounds (22,680 kilograms) of freight.

Such figures do not, however, encompass all of the Navy's assistance, as they do not include all of the operations in southern bases and centres which were supporting the Darwin operations. For example, Kym Yeoward, working in an Army Air Dispatch unit, recalls her time at RAAF Laverton, in Victoria, from the evening of Wednesday 26 December 1974, working around-the-clock loading materials. They packed the freight on 71 RAAF and RNZAF C-130 flights for ten days without stopping.

Johnston summed up:

> The ships' companies of the vessels employed in *Navy Help* could look back with a considerable degree of satisfaction upon their achievements. Seventeen thousand, nine hundred and seventy-nine working days had been worked ashore and a total of 1,593 blocks had been cleared, principally in the northern suburbs. In addition to residential blocks, schools, hotels, commercial and government buildings had had their surrounds cleared and, where possible, repairs effected.
>
> Of particular interest to the residents of Darwin was the clearing and setting to work of the Parap and Nightcliff swimming pools, thus giving access to some recreational facilities within the wrecked city.
>
> In the electrical task skill area considerable assistance had been given to both the PMG and the Electricity Authority in the fields of house rewiring, general repair to exchange and terminal equipment, power line repairs and generator installation in large complexes. Places outside Darwin such as the leprosarium, quarantine station and Gunn Point Prison Farm also were the subject of work by the electrical task skill groups.
>
> In addition, air conditioning in such places as the Darwin Hospital and a number of hotels and motels was also repaired and set to work. Re-roofing and weatherproofing had been carried out on houses in the Rapid Creek area and on priority buildings within the city of Darwin. This latter task included such places as the YMCA, a number of hotels and motels, and the brewery, which I understand rewarded the work teams in kind.

Johnston continued:

> In addition to their transportation of work teams, helicopters were utilised, using both internal and underslung loads, to land the extensive stores and materials required in the repair and reconstruction of Darwin. They also carried out aerial photography and body searches and were heavily involved in surveys being undertaken by personnel from the Department of Housing and Construction. In all, the fleet's helicopters flew 313.3 hours, lifting 7,832 personnel and 244,518 pounds weight of stores and equipment.

The Captain of the *Arrow*'s situation

Meanwhile Lieutenant Dagworthy, captain of the sunken *Arrow*, was pondering his future:

> I stayed on in Darwin with the crew for about a week, assisting with *Navy Help*, and then together with those who wanted to, drove our cars to Alice Springs. There the cars were put on a train, and we were flown back to Sydney or other home ports.
>
> Having lost my ship, having lost two of my ship's company, and being mindful of what happened to the captains of the *Melbourne* after the *Voyager* and the *Frank E Evans* disasters, I believed that my naval career was finished. While in Sydney, I told my wife that I would be court martialled and that I would have to begin a new career. I believed that I would have a blot on my copybook from what happened in Darwin. I thought I would be found to have made mistakes and would be made to accept the consequences.
>
> I remained in Sydney until HMAS *Melbourne* arrived in Darwin and a Board of Inquiry was convened. I flew to Darwin where I stayed with the ship's company at

Coonawarra. We all gave evidence at the Board of Inquiry, which lasted a number of days. Commodore Guy Griffiths, the captain of HMAS *Melbourne*, was the president of the Board of Inquiry.

At the end of the Board of Inquiry, I was called back to HMAS *Melbourne* and was told to meet with Commodore Griffiths in his cabin. I was expecting to be told that I was to be court martialled. In fact, my wife who was in Sydney, had been phoned by one of my year mates, asking when my court martial was to begin.

Commodore Griffiths was at his desk. He handed me a signal and said that these were the recommendations from the Board of Inquiry which would affect me. He had highlighted the sections that he wanted me to read. In part, it stated that Lieutenant Dagworthy should be immediately posted back in command to demonstrate our total confidence in him. I was stunned as I was not expecting that outcome. It took me some time to compose myself.

Commodore Griffiths, later Rear Admiral Griffiths, was so supportive of me at that time. He totally understood what I was going through. In his report, Commodore Griffiths made the observation that if it wasn't for the team spirit and teamwork of the ship's company, far more lives would have been lost on *Arrow*. He used more eloquent words than that, but in essence that is what he stated.

One interesting aspect of Commodore Griffiths' comments directly praised Lieutenant Dagworthy's leadership abilities, for it was under his command that *Arrow* had developed the team spirit and teamwork which saved almost all of the ship's company.

In fact, the Board of Inquiry was specific, saying that "… his display of seamanship during the underway period be commended", and further:

That the ship's company be commended for their devotion to duty throughout the sea phase in the extraordinary conditions prevailing, and also for their fine display of mutual help to gain safety after the ship had struck Stokes Hill Wharf.

… PO [petty officer] Grose and PO Spencer be commended for their efforts in organising the hands to safety from the ship and from the wharf area.

Bob Dagworthy returned to his career, and the next 40 years would see him serving in a variety of posts and steadily promoted. In 1993 he was decorated with the Order of Australia in recognition of service to the RAN and the Australian Defence Force as Defence Adviser, New Delhi.

Organising Transport

The transport situation was central to the tasking of the work parties. Darwin is very spread out, with the "northern suburbs", where most people lived, some ten kilometres and more away from where the warships were berthed. Vehicles would be needed to get work parties around, and many of them would need to be trucks or at least utilities, to get rubbish to the dump. Toz Dadswell recalls organising the situation:

When I drew up the master plan for the clean-up, I was very much aware that we needed many trucks for moving the work force around Darwin. We had to get them from the wharf area to Admiralty House and then after they had been briefed on the day's activities transport was required to move the force to Nightcliff. This problem became very apparent on 2 January, our first full working day. We only had three or four trucks, so a lot of valuable time was lost.

I contacted the local civil organisation and they promised to help as a convoy of trucks from down south was due that day. True to their word, the next day at 0600 about ten or twelve more trucks arrived. We were given full control over the employment of these trucks and the drivers became closely involved with the operation. Many of the truckies adopted a ship and stayed with the same teams up until the fleet sailed.

So, we had our transport but still had another problem. Fuel. There was plenty of petrol in the tanks around the town, but the petrol pumps had been knocked out of action. With the arrival of *Stalwart* the problem was removed. A team from *Stalwart* dismantled the pumps at a service station in Nightcliff, took them back onboard, and within a day returned and reconnected the pumps. We then put a group in charge of the petrol station. They hoisted the White Ensign over the building – an illegal action. However, that kept the locals away for a while. The owner of the station was on side. We kept records of fuel supplied to our trucks and I guess he was paid in due course. So that's the story of HMAS *Petrol Station*.

Having their own transport meant some naval personnel could set up living in places around Darwin, rather than on the ships. Peter Bloomfield, then a senior sailor, recalls:

Our team was made up of about 12-15 technical sailors (shipwrights and a few junior sailor stokers) and we came and went as we wanted thanks to a kind-hearted government official who lent us a four wheel drive and a mini bus. Being on the outer (I think) with the Fleet Shipwright Officer on the *Melbourne*, we were banished to the outer fringes – the Quarantine Station and the Leprosarium[3] ... located out past HMAS *Coonawarra* ... Being your average sailor and suspicious of all things we are not experts on, we were a bit apprehensive about working out there in case we caught the disease and bits of us started falling off. All of the patients had been moved south before Tracy hit. There was not a lot of damage except for the usual bits and pieces torn off, with the main damage to the laundry building – roof missing and a wall down. Our job was to repair the roof and the wall.

The only worker in the laundry was an aboriginal fellow who had both hands bandaged. We avoided him like the plague and stayed about ten feet away from him (leper we thought!). He was very pleasant and would always offer us a cup of tea and we would refuse of course. On about day three he approached us and informed us that he had burned his hands shutting down the laundry during the cyclone and was not a leper! Did we feel like a pack of idiots or what. Sighs of relief, mumbled apologies and introductions all round ensued with many a brew (and the occasional cold beer) enjoyed with good company for the remainder of our time there.

---

3   The disease of leprosy, which is highly contagious, was until recent times dealt with in most countries by isolating those who had contracted it well away from population centres. The Northern Territory, like the other Australian territories and states, had a leprosarium where those who had contracted it were isolated.

WRANS, possibly from HMAS Coonawarra, helped maintain communications ashore during Operation Navy Help Darwin.

WRANS in Darwin

Women were present in the naval service, having been brought in as a concept in WWII, and then disbanded afterwards – and then brought back in again. Naval officer Andrea Argirides wrote in a brief history of their role:

> The addition of women to those manning warships had really never been tried in the world's navies before WWII, and it was too much of a step to see it started then. But their use ashore was too obvious to see it ignored. Although it got off to a slow start, by the end of the war many navies were using women in shore bases in quantity. On 21 April 1941, a Navy Office letter to the Commodore-in-Charge, Sydney, authorised the entry of women into the Australian Navy as the Women's Royal Australian Naval Service (WRANS).

By 1945, a total of 3,122 women had enlisted in the WRANS. Post-war rationalisation led to the service being disbanded and the last wartime WRANS were discharged in 1948. By

1951, however, the need for female sailors and officers was once again recognised and the service was reconstituted. By 1974, they were serving in all sorts of capacities, although it was not until 1984 they were incorporated into the Permanent Naval Forces.

Although the WRANS worked in all sorts of areas, Captain Johnston, ever cheerful, noted in language typical of the times their most visible role:

> May I say in passing, and I hope this will not be taken as a chauvinistic comment, that one of the most delightful sights on Christmas Day and Boxing Day was my WRANS couriers, who had lost the majority of their clothing, riding motor bikes through the city in the skimpiest of bikinis but none the less with their service caps planted firmly on their heads. This unusual garb was to earn me a severe censure from the Fleet Admiral on his subsequent arrival in Darwin, a censure which I took but lightly.

Toz Dadswell recalled the despatch riders:

> After I disembarked from the C-130 in Darwin I went looking for Eric Johnston. He didn't seem all that pleased to see me, but I guessed he was very weary. He told me to set up an office in the airport building to monitor what was going on. So I spent my first night in the make-shift ops room at the airport. Around midnight a young WRANS sailor came through the door. She was a mere slip of a girl and was soaking wet. She turned out to be a "dispatch rider". A group of WRANS who owned motor bikes had taken it upon themselves to be messengers. They collected signals from Shoal Bay and delivered them around Darwin. The weather was dreadful, and the roads strewn with debris so they had a difficult and somewhat dangerous job. I was worried that we might have an accident but as they came under Eric's command, I kept quiet. When Admiral Wells arrived, he and I were driving around the city and one of the girls roared past. The admiral asked what the story was and when he heard he immediately put a stop to the activity and directed that the messengers were to travel by car. The "bikies" didn't agree, and they quit.
>
> I was most impressed with the way the WRANS worked in Darwin after surviving Tracy. They were everywhere, helping out, especially at the airport helping load the families onto aircraft as part of the evacuation scheme. They deserve recognition for their services.

The Navy's Departure

Like its arrival, the departure of the Task Group was staggered. *Balikpapan* and *Flinders* departed early, on 7 and 9 January respectively. The destroyer escort *Stuart*, towing the damaged patrol boat *Attack* to Cairns, sailed in company with *Brunei*, *Tarakan* and *Wewak* on 17 January. The destroyer *Hobart*, the carrier *Melbourne* and Clearance Diving Team One left on 18 January; *Betano* on 23 January; and *Supply* and *Vendetta* on 24 January. The Shore Command Headquarters was closed down on 30 January and FOCAF transferred responsibility for the continuation of disaster relief to the Commandant of the Army's 7th Military District. The following day the last ships, *Brisbane* and *Stalwart*, sailed from Darwin.

The departure of the Task Group did not, however, signify the end of the RAN's support to the rehabilitation of Darwin. In May and June 1975, the minehunters HMAS *Curlew*, HMAS *Ibis* and HMAS *Snipe* surveyed the approaches to Darwin and the harbour itself, locating trawlers sunk during Cyclone Tracy, and other navigational hazards.

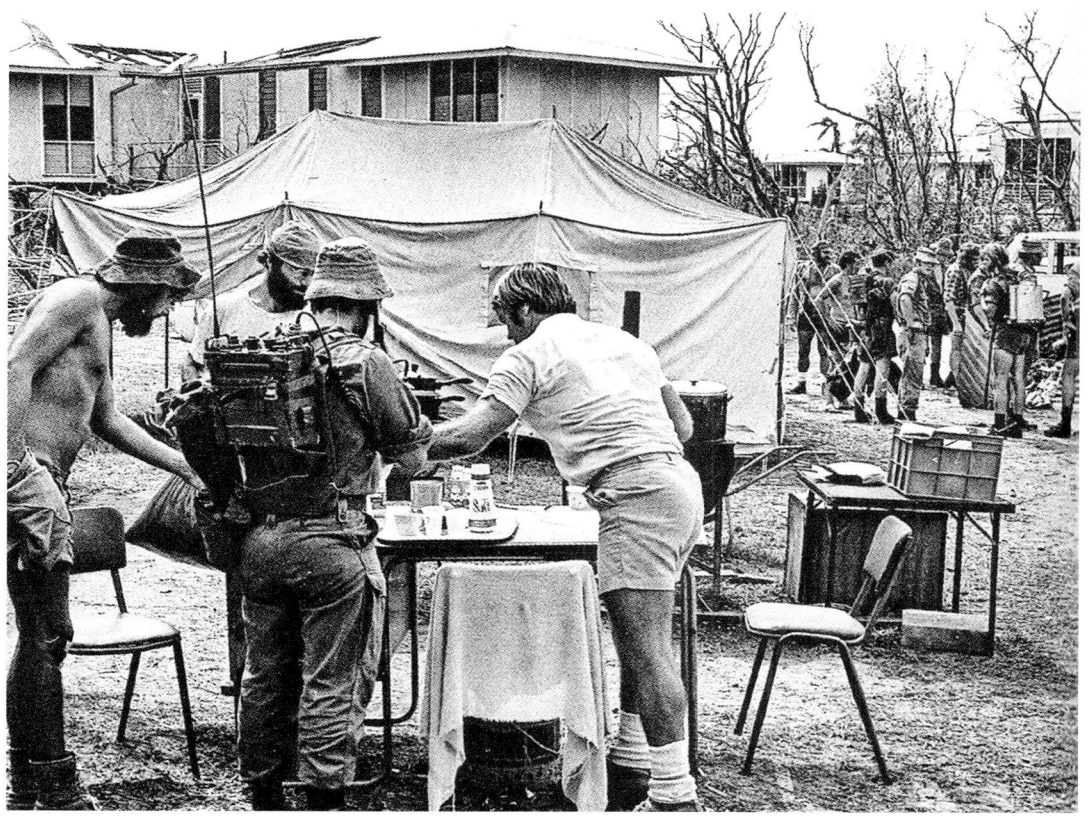

The small Army presence in Darwin helped restore communications immediately after the cyclone. Among these Army personnel is a soldier with a portable radio set.

CHAPTER 6
THE ARMY STORY

The Army was the smallest service present when Tracy struck in terms of personnel, with a strength of 160, made up of essential elements around which further numbers could be built. Despite the initial efforts being carried largely by the Navy and Air Force, in sheer tonnage moved to help Darwin post-disaster the Army comes out on top.

The initial Army presence was commanded by Lieutenant Colonel RB Rogers. The individual units consisted of:

	Officers	Other ranks
Headquarters, 7 Military District	11	47
121 Light Anti-Aircraft Battery and Training Troop		2
125 Signal Squadron	1	13
Detachment – 7 Signal Regiment	2	14
711 Supply Company	2	25
7 Transport and Movement Group	1	19
Darwin Workshop Platoon	1	22
Totals	**18**	**142**

No report has been located which shows the number of Army dependents, but no Army members or dependents were killed or seriously injured in the storm. One Army member, injured enough to warrant specialised medical attention, was taken to Darwin Hospital.

Immediate duties

One officer was ordered to take a measure of the damage the next morning. The Army liaison officer at the RAAF base was released to return to the barracks around 0630. He left his radio set and operator behind and took about an hour to make his way approximately five kilometres between the two bases "driving over roof tops on the road, along footpaths and through people's yards." The damage to the barracks was rated at severe, with nothing habitable, and with "49% of buildings unusable or damaged beyond repair".

Army initially supplied some assistance to policing, without actually taking on police powers themselves:

> … Commander 7th Military District will accept the request. Pers. will not carry arms. There will be one police officer with each group carrying arms. Troops are to assist police in each area in any task they may request.

Army soldiers were used to guard petrol depots, but unarmed, and alongside NT Police members. The Army also provided accommodation for up to 100 policemen but noted that the quality of the living quarters "… will be rough".

One often-overlooked aspect of the armed force's capacities was their organic communications abilities. This began on Christmas Day. General Stretton later wrote:

> Just after midnight, a captain from Army Signals arrived and started to install a radio set in the room we were using in the Darwin Police Station. Within an hour these vital communications were restored.

Stretton's final report noted the extensive communications networks the Army then established:

Army Field Radio

Besides providing radio communications within the military area and with the RAAF Base before and during the cyclone the Army provided the following facilities with manpack sets and operators:

Emergency Command Net [meaning a different frequency for each of the three networks]

This net grew from two radios (one at HQ 7 Military District and one at the RAAF Base) to a third radio at the Police station set up at 1200 on 25 December to a total of five radio stations. The last two were on the MV *Nyanda* and at the evacuee embarkation centre.

Evacuation Net

This net was established at 1600 on 27 December to aid control of the evacuation program. Initially four radio stations were provided for Evacuation Headquarters, the Bus Depot, the airport and Darwin High School. Later three additional radio stations were provided for the high schools at Nightcliff, Wagaman and Nakara. This net operated until 2 January.

Clean Up Net

This net was established on 29 December to aid control of the clean-up of rotting perishables from houses and retail stores. It involved three mobile control stations and twelve mobile outstations with the clean-up parties until 3 January.

In addition:

> By using HF radio from 125 Signals Squadron Darwin and HMA Ships *Assail* and *Advance*, communications to the rest of Australia were possible from mid-morning 25 December.

The Army also started to look to distributing some of their ration packs. They had enough for 350 men for 42 days, but the packs were in a warehouse "which had been completely destroyed and collapsed on top of the ration stacks". On 28 December at 0400 Army trucks joined the Food Distribution Committee work.

Eventually Army's role was to grow enormously. The Navy's operations were to come to an end on 30 January, and the Air Force's evacuation role had been supplanted by chartered civilian aircraft. But thousands of houses still remained to be cleared off their blocks if they were not to be repaired. It was a job the heavy lift capability of the Army was well suited for. Toz Dadswell remembers doing a handover:

> I didn't have any contact with the Army until we were leaving, and an army officer (I think about Lieutenant Colonel rank) came to HQ to see what we had been doing. I took him

Debris strewn house yards immediately after the cyclone, thousands of which needed clearing by Navy and then Army work teams. (RAAF)

out to Nightcliff and showed him the 1,600 blocks we had cleared. That's about 400 a week. All he said was that the Army would do much better as they were moving in with bulldozers and heavy machinery.

Governor-General Cosgrove's experiences in Tracy in his earlier days

One Army member who worked on Cyclone Tracy operations later wrote of his time. Peter Cosgrove, although retired and today one of the most famous faces of the armed forces, had a long and extensive career in the Army. Born on 28 July 1947, he grew up in Paddington, Sydney, and joined the Army in 1965, training as an officer at Duntroon in Canberra, where he recalled he was "disorganised, unpunctual, untidy and immature". Nevertheless, he graduated and served in the Vietnam War where he earned a Military Cross.

Later, he became commander of the International Force for East Timor (INTERFET), the peacekeeping mission in East Timor as it left Indonesia to become a country in its own right. He became Chief of Army; then of the Defence Force, serving with the Army for more than 40 years. He became Governor-General on 28 March 2014 and served until 1 July 2019. But in 1975 as a junior officer, he was deployed to Darwin.

Peter Cosgrove later wrote:

> I recall, as a colleague and I flew into Darwin on an Air Force Hercules, looking out of the aircraft window at a scene that reminded me of photographs I'd seen of Hiroshima after the A-bomb exploded in 1945 … the policy of evacuating all fully dependent folk out of the city and nearby was both sensible and difficult to execute. While the damage caused by Cyclone Tracy was quite local, the implications, the humanitarian aftermath and the urgent and vital clean-up and rebuild were truly national. With the slim population overall of the Northern Territory, it was a case of all hands on deck from the Commonwealth and from states in a position to help.

> It was also an intuitive, logical use of the defence force in a time of national emergency. Major components of the Royal Australian Navy Fleet rushed from their bases into Darwin harbour (although the "rushing" took quite some time – many of the ships came from Sydney). But I know from the locals that the sight of the Navy steaming into the harbour was a huge boost to morale, as was the arrival, a few days later, of a major Army unit (the battalion to which I was posted at the time) on plane after plane of chartered aircraft into Darwin airport.

> The sailors, of course, were not equipped for a great deal of the clean-up work, which was first and foremost among the dire needs. But they were there, highly visible and working very hard, and that was the main point. When the Army arrived from the south in big numbers, we started to get a more comprehensive view of what was needed, and we were there for the longer haul. Essentially, though, our impact was as much on the morale and optimism of the Darwinians and on the sense of concern felt by the wider national community.

> Even though at the time I was pretty junior and entirely focused on leading my company of infantrymen on their allotted tasks, I formed a conclusion that among the myriad needs and wants of disaster-struck people, two items that would always be in short supply were hope and optimism. This, and the insidious phenomenon I have termed "disaster fatigue", were recurring factors in dealing with any natural disaster in the decades of my further experience in this area.

> We were in Darwin for about seven weeks before being relieved by another infantry battalion, 6 RAR, led by my old company commander from early 1969, Tony Hammett. As we flew out on a chartered aeroplane, the soldiers were raucous and overjoyed, knowing that they had performed extremely well, and their presence had made a difference.

A Bailey Bridge

An assessment was made of the possibility of an Army bridge to repair the damaged iron ore wharf, where there was a gap caused when a trawler was smashed through a section. Following an examination and report, a 200-ton 79 metre Bailey Bridge was moved via road convoy from Penrith through Mackay to Townsville, despite floods encountered on the way.

A WWII-era Bailey Bridge under restoration at the Darwin Military Museum in 2022. The bridge had been stored for many years by the NT government before it was donated to the museum. It is likely the same one brought to Darwin by the Army after Cyclone Tracy. (Darwin Military Museum)

There the bridge components were loaded onto one of the Navy's Landing Craft (Heavy), which saw the bridge equipment delivered to Darwin on 13 January.

A US Army soldier's story

One interesting story with a positive spin was the tale of a US Army captain who was on exchange with the Australian force. The *Australian Army* newspaper – distributed free around force bases – found a way to tell the tale of the Bailey Bridge and more in a good light:

A stormy way to breeze through an Aussie tour

United States Army exchange officer Captain Mike Viani, of Boise, Idaho, is a great believer of the old saying about an ill wind blowing nobody any good. He arrived in Australia, last September on exchange to the 1st Field Engineer Regiment at Holsworthy. By January he was in Darwin with the Field Force Group Darwin assisting in the aftermath of Cyclone Tracy.

"This has certainly been the highlight of my tour in Australia," he told *Army*. He came to Darwin with the Engineers primarily to build a Bailey Bridge over the shattered pier leading to Darwin's only deepwater wharf.

Strengthening of the pier to take the Bailey Bridge took the contractors longer than expected, so the Engineers were diverted to repair work around Darwin's cyclone-damaged northern suburbs. Co-ordinated by the Department of Housing and Construction, the Engineers have been weatherproofing the few lightly damaged homes in the area, to provide accommodation for Darwin's returning workforce.

This has involved the repair and replacement of wall panels, roofing timbers and the reroofing of the houses. Specialists have been checking and repairing power and water services to the houses.

The Darwin visit, as well as his experience as OC of the 1st Field Squadron, before coming to Darwin, has given Captain Viani a basis on which to compare the Australian sappers with US Army sappers.

"The Darwin trip has been a real experience, not only from the challenge of building the Bailey Bridge under extremely adverse conditions; but from the fact that the repair programme has given the unit construction training in an area not usually open to Army tradesmen," Captain Viani said.

"As individuals, the general make-up is basically the same. Personal and organisational problems are very similar. The Australian Sapper is highly professional and in general has a higher degree of individual training," he said.

In his short time here, Captain Viani has naturally picked up some of the Australian idiom, and, confidently makes references to CJ Dennis's poems, and the difference between Australian and British accents, but is yet to master the trick of safely opening a can of beer by tapping the top or crimping the side.

An Army Force Field Group of 650 personnel was assembled and moved to Darwin, with a primary responsibility of individual home clearance, the repair of services, safety checks and the re-roofing of homes. It was assembled and in place by 30 January. On that day the Navy formally handed over its tasks to Army.

The Army's long-term task

Army policy envisaged clearing a home site only if the householder were present and, because of the climate, nature of the work and primitive living conditions, the relief of troops would take place every six weeks.

The first Army group was relieved on 4 -7 March by 650 Field Force troops, including 350 Supplementary Reserves from 11 Chief Engineer Works and 21 Construction Regiment for the period of their annual camp, 7-21 March. From the Reserve units, 237 personnel volunteered to extend their annual camp by an additional week. The number of troops engaged was maintained at 650 by replacing the Reserves with Regulars.

The Reserves were most keen, as the Army's newspaper reported:

The continuing story of the relief and reconstruction of Darwin

To put on a uniform for two weeks and then do the same job you have to do in Civvie Street sounds like taking a busman's holiday.

But for 350 Army Supplementary Reservists, this year's annual camp is far from a holiday.

In fact, the soldiers are probably working harder than they do in their normal jobs.

The soldiers, from the Haberfield, 7 NSW-based 21 Const. Regt., are currently working alongside ARA Diggers helping to clean up Darwin.

The Supplementary Reserve engineers are unique. Formed in the 1950s, mainly from State Government employees, they have special skills not always available in ARA engineer units.

The 21 Const. Regt. is made up of men drawn from the NSW Departments of Main Roads; Metropolitan Water Drainage and Sewerage Board; Public Works; Maritime Services Board, and the Snowy Mountains Authority.

About three quarters of the unit live in country areas. Their skills range from civil engineering to plumbers, carpenters, plant operators and even wharf builders.

The CO is Lt.-Col. Bob Allen, who in civil life is a supervising engineer in charge of technical training with the Department of Main Roads.

Valuable Experience

"We have been waiting for this type of training for years," Lt-Col. Allen told *Army*'s Capt. Kevan Wolfe. "It's very valuable as it enables us to function as a Regiment in an operational role. Each of the Regiment's four 7 squadrons are designed to operate independently and this is the first time we have all been able to work together, at the one time in the 24 years that the unit has been formed" he added.

The Regiment has been given the task of rebuilding the staff living accommodation at the Darwin Hospital, which, was all but demolished by Cyclone Tracy.

The Sappers are setting to with a zeal that would make many contractors boggle with amazement, reports Capt. Wolfe.

They are so keen that 238 of them have volunteered to stay an extra week on the job. Some have even taken special leave from their normal employment.

By the end of their stay, the engineers will have completely removed an old concrete building declared unsafe, removed the top floor of another and converted it into a single storey block, re-roofed another and rebuilt one from the ground up.

All to regulation standards ready for immediate occupation.

Tremendous benefit

When asked why tradesmen joined the Supplementary Reserve, Lt.-Col. Allen replied: "The benefits they get from a unit of this type are tremendous. Army methods certainly show up in their civilian jobs, there's the comradeship, too, and they learn something new at an annual camp every year.

A lot of men from the unit have been helped by their Army training, many have developed their leadership potential and have advanced to top jobs".

S-Sgt. Eric Kalio is a foreman with the DMR. For the first few days he was directing the demolition of a dangerous building. He was asked how he felt wrecking a building when he normally builds them: "It's a good break … great for relieving frustrations" he said.

Cpl. David Robbins is a Prisons Officer. He spent his first week in Darwin swinging a sledgehammer and breaking up concrete: "It's hard work – look at my hands – but it makes a change" he said.

By early April the daily rate of block clearance had increased from 70 to 100 and it was expected that the task would be finished before the planned completion date of 9 May.

Although houses were a primary target for the clear-up teams, aiding businesses to get back on their feet was also a priority. For some the prospects were grim: very few people would be buying a new car for a while, and there would little demand for toys, luxury goods, make-up and so on. But other businesses would be in demand: food still needed to be supplied; new furniture would likely be a priority and clothing replacements would be needed by many. The Army assisted with not only cleaning up but with supplying generators on loan, re-roofing shops and taking away the results of building damage.

A platoon commander's story

Lieutenant Peter Pedersen of the Army, later to become a distinguished historian, amongst other achievements, wrote at length of his own command experience in the Tracy clean-up. Published in the *Army Journal* in 1975, it is reproduced here with his permission:

A Platoon Commander's Experience [1]

Lieutenant PA Pedersen

Royal Australian Infantry

On Christmas Eve 1974 Darwin was struck by Cyclone Tracy and the city was largely destroyed by the fury of winds that reached a velocity of 180 miles per hour. Fifty people were killed, and scores injured. Military assistance to the civil community was quickly provided in response to the disaster.

All three services made significant contributions. This article deals with the activities of the Army's Field Force Group which served in Darwin from 18 January to 8 March 1975. This group consisted of over 650 men mostly from 5/7 RAR but also including squadrons from 2 Cav Regt and 1 Fd Engr Regt. This Field Force Group was relieved by a second group of similar size from 6 RAR and the Army Reserve who worked in Darwin until the end of April 1975. Specifically, I shall deal with the experiences of 12 Platoon, D Company, 5/7 RAR.

---

1 Pedersen, Lieutenant PA. "A Platoon Commander's Experience." *Army Journal*. No. 316, September 1975. Used with permission.

Dr Peter Pedersen AM later became a company commander in 5/7 RAR and then its commanding officer. As a soldier/scholar, he was also a leading Australian military historian, writing nine books on the First World War, contributions to several others, and numerous articles on other conflicts. He appeared frequently on Australian television and radio, spoke at many military history conferences in Australia and abroad, and led battlefield tours worldwide, included leading and organising the first British tour to Dien Bien Phu. A graduate of the Australian Command and Staff College and the University of New South Wales, he served as a political/strategic analyst at the Australian Office of National Assessments and as Assistant Director at the Australian War Memorial. During the centenary of the First World War, Dr Pedersen was consultant historian for the Sir John Monash Centre and other Commonwealth commemorative projects on the Western Front and for the ANZAC Museum in Beersheva, Israel, which commemorates the ANZAC campaign in Sinai and Palestine.

The highpoint of a cadet's four years at the Royal Military College is his graduation. Meeting the Governor-General, receiving the Queen's Commission, watching the days-to-go board change to zero – these all go towards making an unforgettable day. If the cadet graduates into the Infantry an even greater highpoint comes when he meets and takes over his platoon. The feeling is one of trepidation. What is it going to be like? Who is my platoon sergeant? What is the daily routine of the battalion? How long will it take to settle in? These thoughts were running through my mind when I marched into 5/7 RAR on 20 January 1975. But things were different from what I had expected. My company was already engaged in the relief work in Darwin following Cyclone Tracy. Two days later, surrounded by the rubble of a ruined city, I joined them.

My initial impression of the effect of the cyclone was much the same as most Australians would have felt. Many buildings were unroofed – many more were completely demolished. This was particularly evident in the northern suburbs of the city – Moil, Nightcliff, Casuarina and Rapid Creek – which were devastated. It is no exaggeration to say that these areas looked like photographs of Hiroshima taken after the bomb had been dropped, such was the extent of the destruction. Iron telegraph poles were bent parallel to the ground and refrigerators tossed about like matchboxes. The wonder is that the death toll was not higher. I was staggered by the magnitude of the damage.

Our quarters in Darwin were at Larrakeyah Barracks with the rest of the Field Force Group. It was a beautiful place, surrounded on three sides by the sea. I remember leaving my room on the first morning I was there and walking out onto the verandah that overlooked the harbour. It was a perfect day, the sun beating down on a sea as smooth as glass. As I stood there, I imagined I could see aircraft – aircraft with red markings under their wings. The date was not 22 January 1975 but 19 February 1942 and around me was not the destruction wrought by Cyclone Tracy, but the havoc inflicted by Japanese bombs. This impression remained with me for the seven weeks that I was in Darwin. It was interesting to talk to people who had experienced the bombing in 1942; to see on the ground where the bombs had actually fallen; to hear how the Zeros would fly low over the city with their undercarriages lowered to attract machine-gun fire. To an observer with an historical bent it was fascinating, akin I think, to visiting an old battlefield.

Back to reality! At about 1145 on that day I met my section commanders. We were having lunch as a company underneath the remains of the Fannie Bay Gaol and my platoon sergeant pulled the section commanders aside and introduced them to me. They had been working all morning and were covered in grime and sweat. By the time we had exchanged a few words it was time to resume work. From then on, we were together. I would work with each section for one day, rotating through the platoon every three days. My sergeant would work with a different section. In this way I got to know my platoon far better and much quicker than would have been the case at Holsworthy. This was one of the great benefits of the Darwin operation and I will have more to say about it later.

What was a typical day in Darwin like? Occasional rest days apart, one day was much the same as any other. We would finish breakfast as close to 0700 as possible and then go over to the company orderly room. This was simplicity itself, consisting only of an FS table and a stores area. We would then confirm our allotted tasks with our company commander and plot these on the CQMS's map. He had to know the locations of our work sites so that

A civilian truck in use by an Army section daubed with the name Miss Tracy.

he could bring fruit juice and ice to us. This was extremely important when one considers the arduous nature of our task and the heat in which it was performed.

Then we would check our platoon lines. A section had its own room, the soldiers sleeping on stretchers. Accommodation was primitive but in post-Tracy Darwin this was normal. These inspections were also important, for dirt meant sickness and sickness might rapidly spread. The need for clean living quarters was paramount. Following my inspection the platoon would do down to the CSM's parade and pick up the stores needed for the day's work.

In the meantime, trucks would start to arrive. The transport compound was located next to the orderly room and by the time we left Darwin it had become one of the landmarks of Larrakeyah. All traces of grass had disappeared from its surface and the daily arrival and departure of 120 trucks had transformed it into a quagmire. Each section had its own truck which assumed the personality and character of the soldiers who rode on it. This was illustrated by the graffiti painted on the side of the vehicles; *F Troop* or *Sunny's Slobs* to *Cuthie's Mob*. One truck had flames painted around the radiator in Flying Tigers fashion. Every conceivable colour was used. "Darwin Here We Come", "Darwin or Bust" and *Fred's Mob* emblazoned its sides. *This Five-Ringed Circus* was quite a colourful affair – some of the artistic creations that emerged would have put *Blue Poles* to shame.

At the conclusion of the CSM's parade soldiers boarded their vehicles and headed off to the task allotted them by their platoon commander. Upon arrival at his particular location the section commander met the homeowner (it was a requirement for those seeking Army help to be present when a clearance team arrived), ascertained what had to be done and then commenced the task. In the early days a team averaged two houses cleared per day but as weeks passed this figure rose to three and occasionally to four. We tried to finish as close to 1600 as possible. This gave us time to return to Larrakeyah, clean up and receive orders for the next day. Stores would be returned, and a fitness inspection conducted by the platoon commander. This was essential, for rashes, tinea and so forth were fostered by the conditions under which we were working.

At about 1630 the platoon commanders would get their orders. The highlight here and the part most eagerly awaited by the soldiers was the score! How had our company gone against the others? What was the Force's total for the day? How many houses had the Engineers roofed? As time went on, we tended to look upon the score as indicating the results of a race. Indeed, it was the more houses that were cleared the sooner Darwin became habitable. On a good day the force could clear over 100 houses. The contribution of our platoon could be as many as 10.

After the platoon commanders gave their orders, the soldiers were stood down. In most cases this meant a nightly pilgrimage to the Koala, one of the few pubs that was open and close to the barracks. It soon became the soldier's pub; indeed, they were employed as bouncers and barmen as well as drinkers. The Koala was famous for its swimming pool and this facility was used to the full. As the night wore on many a soldier took a dip, whether he wanted to or not. After initial protests the manager gave up and the Koala became a way of life in Darwin for the battalion. Needless to say, 5/7 RAR has ensured its financial success for years to come.

This then was a typical day in Darwin. The next day would signal the start of the same routine. It was easy to become bored and this is where the efforts of the section commander to keep his men going became important. Section commanders really earned their keep in Darwin. It was upon them that success or failure depended.

Upon arrival at a job, it was the section commander who met the owner. He made the appreciation that determined the best, most efficient way of approaching the task. He decided on what equipment was needed, on whether to call for additional trucks or front-end loaders.

If the task seemed too big for his team alone it was up to him to call for extra teams. When roving teams were introduced later on it was the section commander who would drive around looking for work.

Each section had its own radio, and this gave him his flexibility. But the section commander's most important task was to keep his men going. It was dull, hard work, and the routine seldom varied. He had to motivate his section, keep them enthusiastic and ensure the job was done properly.

Seeing the section commanders perform under these conditions was also important from my point of view as the platoon commander. I got to know them in a way that I could not have done at Holsworthy, and the converse is true as well. Hence, we understood each other far better than would normally have been the case. This really applied to the whole platoon. You found out why people had the nicknames they did, how they worked under pressure, who had the quick temper, who was popular and who was less so. It was said to me that the seven weeks in Darwin were worth a year at Holsworthy. Looking back, I can now see how true this was.

One of the problems with a MACC operation such as Darwin is that it interrupts a unit's training considerably and calls for a reorganisation of activities planned for the remainder of the year. But for the platoon commander this was more than compensated for by the knowledge he gained of his section commanders and the rapport he achieved with his platoon by constantly working with them. One of the temptations was to take over completely, to go out with a section and run it instead of leaving the section commander in charge. I was guilty of this and indeed the line of distinction is fine. How much latitude did the platoon commander give his section commanders when the method of operation was essentially at section level? Here again the training was invaluable for it was a platoon commander's problem and he had to find the answer. This also extended to soldiers' personal difficulties. Keeping a man's morale and confidence high while enquiries were being made over 2,000 miles on his behalf was not an easy matter.

How was a typical task carried out? In most cases a team would arrive at a house which had been unroofed and whose interior was completely waterlogged. Plaster peeled from the walls. Debris – wall and window frames, smashed stairs and the like – was scattered throughout. Remains of the roof could often be found in the backyard.

The section commander would start by guiding his truck into the yard and clearing it. Strong winds were still blowing in Darwin and debris that could be flung about was quite dangerous. Indeed, we were told that most of the cyclone casualties were injured in this manner. The team would then move inside and clean out the interior of the house. This could be a difficult task for a roof might be hanging precariously or a wall seemingly about to fall. The section had to decide what to clear first and the safest way of doing it. Front-end loaders proved invaluable for lifting large items – complete walls for example – onto the back of a truck. Having been loaded, trucks would dump debris at one of the many tips and then return to the work site where this process would be repeated. Anything from five to ten truckloads were needed to clear the average house.

Techniques were improved during our stay. Vehicles were always in short supply and to overcome this we used truck/front-end loader teams. If these were available, the soldiers cleared a house and threw all the debris into the street. This would be loaded onto the trucks by the loaders which then proceeded to the next job. It was the most efficient method of operation but could only be used on a limited scale because it caused disruption to traffic.

Opportunity tasking was another method introduced. If a team happened to be uncommitted it would seek work. The section commander would ask an owner if he wanted his house cleared. Invariably he did and thus the team and truck were ready and waiting to do the job.

The Darwin operation was not without its humour and I still retain vivid impressions of incidents that had those there laughing a long time after they had occurred. I remember one particular day – it was late in the afternoon and really hot. Tempers were becoming frayed. It seemed that we would not finish the job and teams were progressively called in until there were five teams working on the site.

An inept front-end loader driver who could not lift a thing did little to help matters. Finally, one of the soldiers, who had decided in the meantime that he was an expert on front-end

An ADF work team clears a Darwin block with the assistance of a front-end loader.

loader operations felt that he could show the civilian driver how it was done. He jumped up onto the loader and moved it forward. However, he did not see the jagged water pipe protruding from the ground and the inevitable happened. There was a tremendous explosion as the huge rear tyre blew out and the 'loader settled slowly on its side. Frayed tempers disappeared in an instant as everyone greeted this with great hilarity especially when the soldier concerned looked up as if to say "What's that pipe doing there?"

Although we went to great pains to ensure that clearance operations were conducted in the safest manner possible, one episode would have made the writers of safety instructions wince. One team loaded a wheelbarrow onto their truck (which had no tailgate) upon completion of their task. A soldier sat in the wheelbarrow. The truck had to negotiate a hill. Gravity dictated that the wheelbarrow would go in the opposite direction, and it did. Wheelbarrow and bewildered occupant went sailing out of the back of the truck. Again, this was greeted with great hilarity for the sight of a soldier flying an airborne wheelbarrow is quite rare. Fortunately, he was not hurt.

Several groups of people are worthy of mention. The first of these are the "blowflies", the battalion hygiene squads who were invaluable in making our effort a success. A team would often arrive at a site where a refrigerator full of meat had lain untouched since Christmas Day, or where refuse had become rotten and maggot infested. The smell was overpowering and made working extremely unpleasant. This was a job for the hygiene teams who went in with their gasmasks and cleared out the mess. Theirs was a sterling effort indeed.

The civilian truck drivers became identified with the sections with whom they worked. Initially they were suspicious of our methods and organisation just as we looked askance at their dishevelled appearance and unruly behaviour. But as both groups began to understand each other, firm friendships were formed and a splendid working relationship established. When we left Darwin, they made a presentation to the battalion in a moving ceremony.

The people of Darwin did much to help us help them. They invariably gave us a beer after we had completed a job, invited us to their functions and showed us a real appreciation for our efforts. Despite all they had undergone they still remained cheerful and held their heads high.

Above all there were our soldiers. Without their untiring efforts the relief operation could not have been the success it was. Their task was difficult and monotonous, yet they never flagged. I retain an impression of one of my sections working to clear a house which had been unroofed and from which half the walls were missing as well. It was raining heavily and very cold. I can still see them, freezing and sopping wet, working against the backdrop of an angry black sky. To me this symbolised the whole effort of the Field Force Group in Darwin.

Our time in the city came to an end on 9 March. It had been a unique training ground where I got to know my platoon and they got to know me. During those seven weeks the force cleared over 3,000 houses, numerous flats and several schools. But the

The Darwin suburb of Nightcliff during the clearing process. The streets and some blocks have been cleared but considerable debris remains. The buildings in the upper left are Nightcliff High School. (RAAF)

effort did not end there – it was to be continued by 6 RAR and the Army Reserve who relieved us. While we were justly proud of our contribution, we were none the less glad that our "tour" was about to conclude. Trower Road, Ross Smith Avenue, Fannie Bay, Nightcliff – the household names of the Darwin campaign – were finally behind us.

A detailed inspection of the city and suburbs on 16 April by the Mayor of Darwin and representatives from Army and the Department of Housing and Construction resulted in the unanimous conclusion that the remaining tasks could be completed by 2 May. As a result, the Army presence was progressively reduced and the troops finally withdrawn by that date.

During March 1975, Army expended 9,376 man-days in house clearance and 1,419 man-days in assisting the civil authorities in other ways. Between 1 and 24 April, a total of 6,779 man-days was expended on house site clearance and 46 man-days in assisting the civil authorities. Total vehicle miles travelled during those periods were 17,844 [28,717 kilometres] and 17,852 respectively.

Tracy's impact across the Armed Forces

Tracy had an impact all over Australia. It is important not to forget the role thousands of people in scores of organisations across the country made. Tracy's impact sent a ripple through the armed services that was felt far and wide. Bill Reid, an Australian Army officer, saw a different side of the Army's reaction, from Sydney, where he and his artillery school took in people evacuated out of Darwin. However, there were eventual problems with a some of the evacuees:

> I was a junior officer at the School of Artillery at the time, responsible for training young soldiers, fresh from recruit training, in their basic artillery skills. Every week soldiers would march-in from Kapooka and when we had 36 available, we would start a course. This provided us with a small but essential pool of General Duties soldiers whilst ensuring that trainees spent the minimum possible time at the school. Once they had completed their training the soldiers were posted immediately to regiments, elevated in pay level, and eligible for benefits such as married quarters, rent assistance etc.

> This cycle was disrupted severely for several months by the arrival of the Darwin refugees.

> Trainee soldiers were diverted from their training to provide support: dixie-bashing, setting and cleaning up dining tables, cleaning ablution blocks, exchanging linen, etc. They also gave up their barracks rooms and lived in tents. Since these soldiers were unable to move from the Army's lowest pay scale to a higher one until they had completed their artillery training, this meant that they paid a financial penalty.

> Initially, there was much sympathy for the refugees, and behaviour was tolerated that would not be normally. But as the weeks progressed, and many of the refugees settled into a welfare-funded lethargy and indolence, there was a hardening of attitude, including from the welfare agency workers (government and Red Cross, etc) attached to us.

> When, after a month or so, an appeal was made to help with these daily living tasks so that some soldiers could commence their training, only two or three refugees volunteered.

> I will never forget three low lifes, who spent each day in hotels on The Corso, making loud, smart-aleck comments as my soldiers paraded one morning. An immediate visit from two of my toughest sergeants, with an offer the low lifes could not refuse adjusted their attitude.

When, to our great relief, the refugees finally left, my soldiers had to clean the barracks before reoccupying them. We were disgusted at their state: soiled nappies in drawers and wardrobes, graffiti and other disgusting marks on walls, cigarette butts and other detritus everywhere, etc.

The only formal recognition for this effort, that I am aware of, was an OAM to the Warrant Officer Caterer, the late Dudley Pye. He and his cooks did a wonderful job feeding the refugees, including preparing dishes that suited their ethnic and cultural tastes. This OAM also recognised his, and his cooks' efforts a few months later when they were called on, with virtually no warning, to feed Vietnamese orphans who were being cared for at the nearby Quarantine Station (Operation *Babylift*).

The second-in-command of the school, the magnificently named Major Donald Macleod Yeomans Smith (now deceased), who also had a memorable handle-bar moustache, was directly responsible for the refugee support. Don Smith worked tirelessly: I hope that his officer's confidential report that year reflected what he did.

Other school staff, such as the Quarter Master and his team, and, of course, the cooks, also deserve honourable mention.

The Army's time eventually came to an end. But it would take several more years before Darwin was approaching anything close to the normality of life as it had been before the cyclone. It had been a tremendous effort from the soldiers. But in later years there was a shadow cast over their activities immediately following the Christmas Day impact. Ironically it was caused by General Stretton's memoirs: he alleged that his own service, the Army, had not done enough. This is discussed only in the Appendix, because in the end it seems to be more of an argument than facts.

ADF bases across Australia were used to house and support Darwin refugees. These Tracy survivors are enjoying a delayed Christmas Party at RAAF Laverton in Victoria. (RAAF)

An armed NT policeman checks for looters in the days after the cyclone.

CHAPTER 7
UNIFORMS APART FROM THE ADF

NT Police operations

Major General Stretton wrote of seeing police immediately at work:

> In his devastated police station, amid the corpses and the stunned women and children, I found Commissioner Bill McLaren at his post with his officers and men. Despite the great personal tragedy and loss that involved all of them, the whole force rallied to the call of their commissioner and reported for duty. I doubt in all my years of service whether I have seen such devotion to duty as that shown by the commissioner and men of the Northern Territory Police Force after Cyclone Tracy.

The local police were soon at work on Christmas morning. Their commissioner, Bill McLaren, had crawled out of the wreckage of his own house and went to work. His family: his wife; a university age son home for the holidays with a friend, and a younger son, had all survived the night with him. An interviewer later asked the commissioner the questions in bold:

Could you describe your actions first thing after the cyclone started to abate on that morning, Christmas morning?

We went downstairs and attempted to go to work. But before I could, all the trees were across the driveway within our block and I had to get the axe and cut all the trees up to be able to shift them off the line. The electric wires and so forth were tangled amongst the trees which had come down near the front gate – we had to prop them up out of the road.

And eventually, when we got that cleared, I then was going to work, and the departmental car that I had at home wouldn't start. So I took our own car; it started and I took it down. It was very badly chipped. It was chipped everywhere from the gravel hitting against it – really left it sort of mottled, particularly on one side.

And then I made my way down to the police station. And there was rubbish and parts of houses and timber and wires, all sorts of things all over the road. And it was a very slow trip down. And [you] had to be very careful as to where you went because of the material that you had to cross over.

Could you tell me the exact location of the headquarters that you were heading for please?

The headquarters was on the corner of Mitchell Street and Bennett Street. And it had only been built – oh, within the last four or five years.

So when you eventually got there, what was the scene that greeted you?

Well, all the way down the street, houses and various buildings had been damaged greatly. At the police station, some of the roof was missing. There was water through everything, or practically through everything. We had an emergency lighting plant; it was operating. In fact, this was one of the few plants in Darwin that had electricity at that stage. The communications section, which was our main lifeline, it was badly damaged and at that stage, there were no communications.

So could you just go over for me, to the best of your recollection, the events of that first day?

Well, we had a look round and tried to get an idea of the extent of the damage from around Darwin, from some of the members who went out on patrol and returned with their various reports. And then when we had some idea of the extent of it, we held a meeting of all the officers.

The biggest percentage of our members had turned up for duty, which was rather remarkable because they had suffered damage the same as most other people. In fact, many of them had completely lost their homes, and yet the biggest percentage of them turned up. There were some who didn't but then on the other hand, the ones who didn't, I think did quite a lot of work in the area where they were, or had been, living.

After we got these reports of the extent of the damage, we held a meeting with all the officers who were present. And in accordance with our previous plan, allotted the various duties to them. For instance – someone to be in charge of transport; someone to be in charge of communications; someone to be in charge of the injured and to take care of the injured; and someone to look after the – from the mortuary side of it – to look after the bodies. And various jobs were allotted under these particular officers.

The NT Police's presence was understood and accepted by those locals who were staying on. Things were much less formal. Former NT Police member Daryl Manzie recalled:

We had our showers under a concrete mixer with its motor in reverse so the drum poured out water. We just stood in Austen Lane and soaped up.

The police seemingly supplied their own food assisted by their own families. Daryl remembered:

… we were all eating in circumstances that could best be described as a cross between a giant mess hall and soup kitchen. The largest was underneath the Darwin High School, and most police were eating in a kitchen set up at the entrance of the then Police HQ, and run by Eileen Cousins the wife of Inspector Len Cousins.

NT police manning phones in Darwin.

But it became immediately apparent that with an entire police force of 208 members in the NT this would not be sufficient. The prospects of looting and theft were a distinct possibility, and it soon became apparent that there would be a considerable number of fatalities to deal with.

The small NT force was also spread across the entire Territory. Alice Springs was 1,400 kilometres away, the second largest city in the NT, but police were still needed there for everyday matters. This was also the case with Katherine, some 300 kilometres south of Darwin. In addition, the myriad of tiny one-man stations scattered across the Territory could not simply be abandoned. It was immediately apparent that interstate police assistance would be required.

Interstate Police Support

As it was, people down south had thought to pick up the phone. Bill McLaren continues:

> On Christmas Day - it was one of the freaks of the telephone network – I had a call from the Commissioner of Police Reg Jackson in Victoria, and on that day he offered the services of police and I accepted them.

> **Were there any conditions made on the police coming up here, in terms of whether they'd be armed or what their powers would be?**

> No arrangements whatever. They came here as police to assist the police here.

Eventually 162 police personnel came from almost all of the Australian states and territories as per the table below:

Force	Number	Arrived	Notes
Queensland	12	27 Dec 1973	
Northern Territory	208	n/a	n/a
Victoria	30		21+D4 section
		1 Inspector, 1 Senior Sergeant, 5 Sergeants and 14 lower ranking members (Senior Constables and Constables) were sent to Darwin from the 30th of December 1974 for duties including traffic control and general duties.	
New South Wales	48		
Australian Capital Territory	49		
Western Australia	1		No records located by the Western Australia Police Historical Society, except for one forensics officer
South Australia	22		
Tasmania	Nil		

Anyone who lives for some time in the Top End gets to know that "things are different" there. And that was still the case. What were often known as "blow-ins" from "down south" were not usually appreciated, but it appears the deployment of the southern officers worked:

> **There have been some conflicting reports about police. On the one hand the Northern Territory police seemed to have been viewed in a very positive way, whereas the New South Wales police seem to come up for a lot of adverse mention. What was your view of what was happening at the time?**

> Well, I think that really, all of the police did a good job while they were here. But the Territory is a peculiar place, as probably you are well aware yourself at this stage. The

local police understood the local people and understood local conditions. And they had a much more – and they still have – a much more personal contact with the public than the police do in the bigger cities – for instance Melbourne, Sydney, and so forth. And the methods that they had of dealing with certain matters from some of the other states, were not probably on the friendly, understanding conditions or manner in which it was handled by the local police.

And I think – my impression was the local police got tremendous praise for the work they did. They did do a marvellous job and instead of being at home, looking after their own affairs, they came to work and they gave a tremendous amount – they made a tremendous effort.

They did receive a better public image than most of the interstate police. But as I said, they all had a job to do while they were they were here – some of them were unpleasant, some of them they hadn't been used to – and they all had a job to do.

For instance, one of the things that didn't help the police image much at one stage was, there was a tremendous number of animals that had been abandoned as a result of the cyclone; whether they were abandoned or whether they ran away and came back, or whatever it was. And at one stage, dogs for instance, were hunting in packs and it was a case of having to destroy them. So we had to send shooting parties to destroy the cats and dogs, and that didn't go over well with the public but it was something that had to be done.

The interstate police arrived piecemeal – the table above sums it up; went to work successfully,

An armed Commonwealth police officer from the Australian Capital Territory maintains law and order at the airport during the evacuations. (MAGNT)

and over the next months as the city very slowly returned to normal were returned south. Bill McLaren later summed up the interstate involvement:

> Some of them left within the first week – the majority of them left about the eleventh of January. And a group stayed on from Victoria to assist us to re-establish a communications section. And some members of the Commonwealth police stayed on until about the end of January, to allow a number of our people to take a fortnight's – either a week or a fortnight's – rest and recreation leave.

However, stories about looting and other lawlessness made their way quickly into southern newspapers. On one occasion, General Stretton was contacted by media asking if there was "any truth in a Sydney newspaper headline that nine policemen had been shot by looters?"

Stretton at one stage lists the stories "down south" about the situation in Darwin:

> … police shootings, the outbreak of disease, uncontrolled groups of looters taking over Darwin, the concealing of bodies to keep the death count low, [which] were all counteracting the responsible reporting coming from the press corps on the spot.

But law and order was not too great a problem. After the lurid stories, true or exaggerated, made it into the southern media questions were asked in the federal parliament. Later in the House of Representatives the assembled members were advised:

> … there were only 31 arrests during the 2 weeks ending 17 January 1975.

The Red Cross

The Red Cross also were of considerable assistance.

> Early on the morning of 26 December, and at the request of the Natural Disasters Organisation, an emergency team of twenty highly experienced officers was sent from the New South Wales division. Four tracing and registration experts were sent from Western Australia, and Dr Beal, director of the South Australian blood transfusion service, arrived from Adelaide. This initial group was later replaced by a team of sixteen people recruited from the Victorian, South Australian and Western Australian divisions, led by SG Goddard, Victoria's general secretary.

> In Sydney, eighty-two Vas [sic] maintained a 24 hour roster at Mascot airport to meet the evacuees. They acted as escorts on ambulances travelling from the airport to local hospitals with the sick and injured. Branches in Katherine, Tennant Creek and Alice Springs helped thousands of refugees who streamed south in the aftermath of Tracy. Almost 6,000 evacuees arrived in Tennant Creek between 26 and 31 December. They were accommodated, fed and supplied with essential clothing and supplies such as food, nappies and babies' bottles.

> Katherine "swarmed with Darwin refugees", who arrived around the clock, "shocked, depressed and weary". President of the Katherine branch, Helen Murphy, was awarded the Meritorious Service Medal for her leadership during the crisis … The Red Cross disaster team worked seven days a week from 8am until 8pm.

The Salvation Army

Another uniformed presence, the Salvation Army, was there on the streets immediately post-Tracy, helping people both in the stricken city and across Australia. One of them was Ron Wilson, in later years to become a national television and radio news presenter. When Tracy struck he was in his family house with his parents and brother. They lost their house. He recalled the next day:

"Within 24 hours of Tracy's havoc, the Salvation Army was on the streets," Ron remembers.

"Simple things; but things we needed right then and there … food, clothes, a helping hand … a smile and encouragement to keep going" he says.

"The Salvos stuck it out with us in the miserable conditions with no electricity, no running water, no sewerage facilities, in the oppressively humid heat and drizzling rain in the weeks that followed."

Within two months, Ron and his mother were on an evacuation flight to Sydney. His father remained in Darwin to help rebuild, while his brother had left for Canberra and university enrolment.

The only possessions Ron and his mother had on arrival in Sydney were the clothes they were wearing and a few more in a small bag.

"As ever, the Salvos were at Sydney Airport when we touched down," Ron says. "We spent the next six weeks at the East Hills migrant hostel. It was a very frustrating time – no money of our own, no job, no friends; just bad memories.

"The Salvos seemed well aware of the risk of depression and through their daily visits encouraged us on all of the positives that lay ahead.

"Salvation Army officers [pastors] visited us every day, without fail. Each day they handed us a $20 bill. It doesn't sound like much today, but it was enough for treats like a bar of chocolate.

"It also allowed me to work a nearby public phone into overtime trying to track down a job."

Ron was successful in finding a radio job at Wollongong station 2WL (now WaveFM).

"A week later, the Salvos arranged for mum and me to travel to Wollongong, where we were met by a Salvation Army officer who had rented us a unit and paid the first month's rent.

"The unit building was brand new. It was probably better accommodation than I had ever lived in before. Not only that, but it was furnished – all brand-new furniture. Never once was I asked to fill out a form, prove my need or given a lecture.

"It was just friendship and help. Eventually, we rebuilt our family and our lives. And that is why I will always do whatever, whenever, for the Salvos."

Ron has since had a highly successful media career, including a 30-year association with TEN as a news presenter on the channel's national breakfast program and evening news, and presenting news broadcasts on several Sydney radio stations.

The Salvation Army continued their good work at southern destinations too. RAAF officer Ian Frame, flying in C-130s, recalls:

The Salvation Army deserve a special mention. At every destination I flew evacuees, at all times of day or night (e.g. 0200 at Brisbane) they were on hand with food, comfort and, if necessary, clothing, etc for the evacuees. Some other larger agencies were notable by their appearance only in daylight and/or when TV cameras were present.

Reporting is sparse on the activities of the ambulance and fire services in Darwin, but it seems they stayed in place, and like so many workers, provided their essential services.

Members of the Salvation Army oversee the distribution of second-hand clothing to newly arrived Darwin refugees in a southern capital city.

Red Cross workers assist babies being evacuated from Darwin.

Stretton on the balcony of the Travelodge Hotel, probably in his final days in Darwin. (NT Library)

CHAPTER 8
STRETTON DEPARTS

Darwin within days of the cyclone saw some return to reality with politics interfering. At 1800 on 27 December, Stretton hosted a meeting with six federal cabinet ministers who had flown in from Canberra. It was held in the "police operations room", into which 50 people were crammed, while others watched on through "the glass walls".

There was, somewhat unbelievably, efforts made by members of the federal government to take as much credit for the relief work as possible. Six cabinet ministers flew in, creating leadership tensions without adding much to the relief efforts. Stretton tells of one part of the meeting when the Minister for Housing and Construction was said by Senator Lionel Murphy, there present, to have given "personal approval to all the relief supplies that were loaded on to the [naval] Fleet". The minister, however, was speaking at the same time as a long telex paper on the table showed the supplies had already been listed by the Navy.

It must have been extremely frustrating to deal with such matters at the same time as giving attention to essentials such as whether the sewage plants were operating – they were; or whether water was back on – it wasn't. But the pace kept up for Stretton, who spent his days travelling from committee to committee or sometimes being involved in more practical matters – at one stage he was caught up in a legal storm involving the charging of a man for theft.

However, within a week of his arrival General Stretton thought that enough had been done insofar as his role was concerned. Prime Minister Whitlam had announced on radio the formation of the Darwin Reconstruction Committee, and in the same broadcast paid tribute to the National Disasters Organisation and said that Stretton would hand over his responsibilities on 2 January. The Governor-General, Sir John Kerr, would then be flying in.

Overall, the shattered remnants of the Territory infrastructure were coming together. The water was back on, electricity was being restored, the clean-up was proceeding, 80% of roads were open, the *NT News* was back in production and ABC radio was functioning. Dr Ella Stack, then a general practice doctor, and later mayor of Darwin, operated:

> … a makeshift clinic out of Darwin High School, home to some of the few major buildings left standing.

> "Not only did I look after the people that came in, but also people came and lived here" she said. "They brought their sodden old mattresses with them … I used to do a ward round every day and call them the sodden mattress lot."

Stretton attended a meeting held by the administrator to discuss the handover. There was an argument about where the handover should be – in town or at the airport; the latter was decided on. An eleven-point program was drawn up, which Stretton noted he found "embarrassing" largely because he had no actual legal powers to "hand over" to the governor-general as was being implied. He later sent a message south bringing forward his departure time:

To: Acting Prime Minister

Minister for Defence

Minister for the Northern Territory

Information NEOC Canberra

The situation in Darwin is now so good that it is necessary for official administration to take over as soon as possible. The evacuation has been completed and essential services have been restored to an acceptable level. The first elements of the Fleet have arrived. I am happy to report that there has been no loss of life from administrative causes since the cyclone.

I feel my mission is complete and that I am redundant here. My further presence now causing unnecessary delays in the resumption of normal administration and raises a special problem of protocol at the ceremony planned at 0930 2 January.

In any case my presence at that ceremony may detract from the performance of the citizens who have done all the work with the minimum of assistance from me. With the concurrence of the Acting Prime Minister, I handed the following despatch to His Honour the Administrator at 1pm local time today ...

With the concurrence of the Acting Prime Minister, I and my staff officer will be departing for Canberra sometime today in an RAAF aircraft, as soon as one becomes available.

The general went to the local ABC radio studios to join the administrator and the chief of police. There were some mutual thanks made on air. It was all quite low key, and Stretton must have been exhausted. But he did not forget to thank his right-hand man. He later wrote:

After the press conference, I thanked Frank Thorogood [his personal aide] for the outstanding contribution he had made to the relief of Darwin. He had been by my side continuously since we had flown out on Christmas Day. He had displayed a rare devotion to duty and had been an inspiration to all with whom he came into contact. I was disappointed that the Army did not see fit to support my recommendation that he receive formal recognition.

Later than night Stretton flew out on a Hercules. A small amount of controversy followed him, and it has been noted in the Appendix. But he was the right man in the right place at the right time.

Bill McLaren thought of Stretton "I think he was a great one at making broadcasts ... and I think to get the message across to the people of Darwin." Ray McHenry said:

We were fortunate in having an army officer to organise the services clean-up side where the people could go and say, "Look, could somebody come and do it?"

Soon Major General Stretton was given one of the first decorations awarded under the new national honours system and was made an Officer of the Order of Australia. He was also given the national title of Australian of the Year, and that of Father of the Year by the New South Wales government. They were fitting tributes indeed.

Major General Alan Stretton (right) receives the Australian of the Year Award from Prime Minister Gough Whitlam in 1975.

Cyclone recovery efforts continued for years after the initial ADF response overseen by Stretton. This is a meeting of the Darwin Reconstruction Committee on 16 December 1977. (MAGNT)

CHAPTER 9
WHY NO MEDAL FOR CYCLONE TRACY?

The National Emergency Medal, which was established in 2011.

As the 50th anniversary of the most severe domestic disaster to impact Australia approaches at the end of 2024, many are puzzled as to why the armed forces members who rebuilt Darwin after Cyclone Tracy were not rewarded with a medal. For that matter, why has a medal not been awarded to police members, those in the other emergency services who stayed on in Darwin, and those involved in all aspects of saving a capital city from ruin?

Decades after the event, the National Emergency Medal, established in 2011, has been recently rightly given to ADF members, and indeed many civilians who worked through cyclones, bushfires and floods, now going back to 2009. But those who gave their all after the cyclone that killed 66 Australians may not see the anniversary with more than private pride. Yet this was the biggest operation the armed forces of the country ever mounted outside warfare. It was not only in numbers of personnel – around 8,800 – who worked in the northern capital, but in time too: deployments were in months, and the hours put in were often twelve or more per day, especially in the early weeks; and the conditions endured in a tropical monsoon season were often appalling.

Questions as to why there was no overall decoration awarded for this service were soon raised in the media. But there was no action taken, and therefore no group decoration made. There have been suggestions made to rectify this situation, but they have either met with silence or refusal due to various objections. In many ways the lack of action can be summarised as follows:

1. *There is an expectation in the armed forces that unusual or arduous duty is part of the job.* Indeed, rather than be paid an extra hourly rate or similar for work performed after normal 9-5 or weekday hours, ADF personnel are paid what was known then and is now as a "Service Allowance". In 2023 that was $13,448 per year. More money is paid to naval personnel who are at sea, and if deployed to a combat zone ADF members receive additional pay, with income tax suspended for their time. So in many ways it is an expectation that service life demands more. Then again, deployment to relief operations for Cyclone Tracy demanded *much* more: highly unpleasant work in extremely arduous conditions for week after week.

2. *There was no such honour as the National Emergency Medal at the time.* However, there is nothing to stop the federal government making the decoration retrospective. Indeed, this has already been done: the medal was established in 2011 but recognises operations from 2009.

3. *Many people argue that making such a medal retrospective would result in a flood of applications for it to be applied to other situations.* To which must be said, what of it? Why not reward people for service beyond the norm? It can only serve to make

serving in the forces and the emergency services more attractive to both those who served then, and those who serve now.

4.   *There would be difficulty finding the names of the personnel who served in Tracy operations.* This is not a realistic argument. Considerable records exist from the time, and aircraft manifests and ship schemes of complement and the like are plentiful, as are pay records. Each branch of the ADF also contains personnel in historical sections who might well be tasked to find such records.

A summary of armed forces numbers in Cyclone Tracy relief operations:

Force	Number	Arrival	Departure
Navy			
Darwin-based	696	n/a	n/a
Deployed ships	3,000	27 Dec	30 Jan 1975
Air Force			
Darwin-based	670	n/a	n/a
Deployed personnel	3,000	25 Dec	2 May – presumed for final Army withdrawal
Army			
Darwin-based	160	n/a	n/a
Deployed personnel	650	30 Jan	4-7 April
Deployed personnel	650	4-7 April	2 May
Total ADF personnel	**8,826**		

Notes:

•   To be included in the numbers, personnel had to be physically in Darwin at least once.

•   Numbers are approximate – the Navy and RAAF reports use the term "about" when reporting deployed personnel. Army numbers are exact.

•   The Navy deployed the highest number of personnel at nearly 4,000, but the majority of them were present only from 27 December to 30 January.

•   The Army, with its two 650-man deployments, was present for the longest amount of time.

If we added the police, Red Cross and Salvation Army numbers to the ADF total, and an assessment was made of other uniformed personnel such as ambulance and fire services, the numbers concerned would seem to be around 10,000.

There would also seem to be a valid reason for adding those workers who remained in the shattered city, and those who were admitted for their skills in the months remaining. A permit system operated, and only those needed were brought in. The admittance system was an organised one, so records remain of that. Perhaps another 10,000 were brought in to rebuild Darwin. Why not present the medal to them too?

In conclusion, the award of the National Emergency Medal would seem both possible and have much to recommend it for uniformed personnel involved in Cyclone Tracy.

On 28 December 1974 police found three cars packed with looted goods including these life jackets and folding chairs.

CHAPTER 10
TRACY CONTROVERSIES

There were a number of controversial stories concerning Cyclone Tracy. It is not the intention of this work to address all of them, but two need some comment, as they involved, or were said to have involved, members of the defence forces. The first is allegations of theft, and the second of conspiracy in concealing the numbers of fatalities.

The Looting Controversy

There were certainly examples of theft. Author Sophie Cunningham commented:

> There is no doubt some theft occurred after Tracy. Lists presented to the Supreme Court in the wake of various arrests include multiple TV sets, cartons of pantyhose and what now seems like an amusing surfeit of banana lounges.

> But two weeks after the cyclone, only fifteen people had been arrested on charges of larceny and possession of stolen goods. More significantly, the culprits identified in the most persistent rumours were not taken into custody at all.[1]

The Queensland Police Museum carries the following story:

> On Saturday 28 December 1974, Constable First Class Bob Latter, along with three NT Detectives, went to Millner Street in Darwin. There they found three cars packed with goods looted from destroyed homes and shops. The final haul of confiscated goods was huge and included, amongst other things:

> > … 24 rolls toilet paper; 12 pkts razor blades; 12 nappies; 10 striped tea towels; 10 tins of crab meat; 8 folding chairs; 8 pairs nylon socks; 6 floor mats; 6 pairs of blue jeans; 5 bath towels; 5 mixing bowls; 4 teddy bears; 3 size 14 dresses; 3 lace table cloths; 3 pink brunch coats; 2 pkts glazed fruit; 1 cake plate; 1 radio cassette; 1 purple bedspread; 1 Sharp calculator; 1 fishing net; 1 coffee set; 1 aluminium boat; 1 esky; 1 tool box and 1 pair long trousers.

> Several men were arrested for stealing these items. Unfortunately, this was not an isolated case of looting in days following Cyclone Tracy. Police officers assigned to anti-looting duty were kept continuously busy.[2]

One ex-Navy member, then working for the Department of Civil Aviation, and driving south as an evacuee, saw a truck full of loot at Pine Creek south of Darwin, also heading south:

> Sergeant Graham Bowning came up and said hullo, I had known him since he joined the police and was stationed in Alice Springs when I lived there. Graham noticed a 7-ton truck pulling in covered with a tarpaulin, he got up and walked over to the truck, pulling back the tarp we saw beds, washing machines, kid's toys, fridges and whatever. Graham arrested the driver and took him over to the police station.[3]

---

1   Cunningham, Sophie "Descended upon by looters." *Overland Magazine*. https://overland.org.au/previous-issues/issue-209/feature-sophie-cunningham/ Issue 209, Summer 2012.

2   Queensland Government. "From The Vault – Help at Hand… Cyclone Tracy – The Queensland Police Contingent". https://mypolice.qld.gov.au/museum/2013/08/20/from-the-vault-help-at-hand-cyclone-tracy-the-queensland-police-contingent/ 20 Aug 2013. Accessed June 2021.

3   Royal Australian Navy Communications Forum. Unnamed ex-RAN member. "My Cyclone Tracy Experience." https://www.rancba.org.au/Cyclone%20Tracy%20by%20Kev%20Ruwoldt.pdf Accessed June 2022.

A warning to looters outside a Darwin shop.

The strangest story of theft, however, has the Navy as aiding and abetting the-supposed theft of a souvenir of the cyclone from the people of Darwin.

Resident Dee Slater, who later published her story in a book, on the morning following the cyclone, saw a twisted mass of metal about the size of a "commercial chest freezer". Several Darwin residents were looking at it and saying it should be preserved as a sculpture to show the strength of the cyclone, but then:

> … amongst all of the heated discussions as to who actually owned it, two plain-clothed Navy officers then stepped forward. On behalf of the "powers that be" this time they flashed their ID's, then quickly informed us that the sculpture was never going to be for sale. Not at any price. All offers would be refused. Inexplicably, that "monument" has not been seen or heard of since.

The reason for this is that:

> I'm sure that anyone who saw that sculpture today, would have no doubt in their mind (absolutely no doubt whatsoever), that the true facts about the destructive power of Tracy have never been fully revealed to the public. Quite the opposite![4]

However, and the author speaks as a retired navy officer himself, there would seem to be little reason for two such people to be on any duty connected with such an incident in the days following Tracy. Nor do such officers have any legal powers in such a situation, and commissioned personnel in the ADF know full-well what their powers are. The purpose behind their reported intervention is incomprehensible.

Indeed, there *are* two memorials of twisted girders to Tracy. The website *Monument Australia* explains:

> The memorial constructed with twisted girders from the home of Sergeant Kevin Malley which was destroyed in the cyclone in 1974, commemorates those who died in Cyclone Tracy.

---

4   Slater, Dee. *Tracy's Fury*. Self-published: Warwick, Qld; 2018. (p. 63)

At the height of Cyclone Tracy the Malley home totally disintegrated, blowing Malley, and his two children, who were in his arms, fifteen metres to the ground. His wife Joyce fell beside him. He threw himself over his children to protect them, receiving severe injuries which required 200 stitches from flying debris. Joyce had a broken spine and Fiona severe leg injuries which required 23 operations to correct. Only their son Stephen escaped unscathed.

The Cyclone Tracy memorial featuring twisted girders at Casuarina Senior College. (Monuments Australia)

The family was the first evacuated by aircraft to Sydney. The girders were bent by front end loaders during the cleanup operations. A teacher from Casuarina High school took the initiative to use the twisted girders as a monument. The girders were set in concrete and the memorial was unveiled by the Administrator of the Northern Territory in June 1984.[5]

The memorial is in the grounds of Casuarina Senior College in Darwin's northern suburbs. However, that's a fifteen-minute drive from the central business district of Darwin, which is where cruise liners dock and most of the tourist attractions are. Indeed, the monument would be better placed on Darwin's Esplanade, which is where there are numerous other memorials and plaques, mostly commemorating WWII military units. A plaque listing the names of those who died, established by Darwin City Council, is at the time of writing on the wall outside the front doors of the Darwin Council building.

Another memorial, says Sophie Cunningham, is across the harbour near the Mandorah Hotel.[6] This is even more inaccessible, as to see it requires a ferry ride of around thirty minutes across Darwin Harbour.

The Fatalities Controversy

There was much cause for grief in the days following Tracy as death and injury was encountered. The fatalities within a Navy family have already been noted:

> Another terribly sad one I remember was a fellow from the naval base; his wife and two young children. I remember him being there identifying them, and he was just God, the effect it had on him, he was just a shattered man ... Bullock is referring to Geoffrey William Stephenson of HMAS *Coonawarra*, who arrived at the morgue to identify his

5   *Monument Australia.* Website. https://monumentaustralia.org.au/themes/disaster/storms/display/80258-cyclone-tracy-memorial

6   Cunningham, Sophie. *Warning: the story of Cyclone Tracy.* Victoria: Text Publishing, 2014. (p. 250) See also a reference to the Casuarina memorial on page 253.

wife, Cherry Leona Rose Stephenson, twenty-two, his three-year-old stepson, Kenneth James Scott Wheatley, and his six-month-old daughter, Kylie Jane Stephenson.[7]

But apart from the tragic reality, allegations soon began that the count was much higher, and it was being concealed. A further macabre allegation was that bodies were being stored indiscriminately in informal locations. Given the armed forces were very much on the forefront of digging through the rubble; occasionally uncovering bodies, and the whole situation was being managed by an army officer, those stories are addressed here. Note that footnotes have been used extensively in this chapter, in case readers wish to follow the analysis behind giving an accurate fatality count.

The morgue allegation

Dee Slater, the resident who we met above, wrote later she was at a gathering at 0800 on Christmas Day where a "Civil Defence" spokesman was speaking to a crowd of locals:

> … he said that the known death toll was inestimable because bodies were piling up all over Darwin. At 8am over 50 bodies were already lining the long hallway of the main police station in Smith Street alone. The morgue was filled to overflowing … we were then told that all of the post offices across Darwin were now designated morgues (due to their emergency backup power), but unfortunately, they too were filled to maximum capacity. Bodies were piled up against every wall, and on top of sorting tables at every post office. The numbers were staggering …! He then named five of the closest post offices where they would definitely not take any more bodies. Casuarina, with 60 bodies on their mail sorting table alone, was the sixth.[8]

This sounds a bit odd given the time – 0800 on Christmas morning. This was mere hours after the final winds had died away. Indeed, daylight in Northern Territory on 25 December does not arrive in the early hours as it does in summer in the southern states – it only becomes light around 0630. It was also raining still, and the roads were blocked – previously it was described how the army liaison officer at the RAAF base took about an hour on Christmas Day to make his way five kilometres between the two bases "driving over roof tops on the road, along footpaths and through people's yards".[9] How could "50 bodies" have been found and recovered by then? So, it is unlikely the event described by Slater would have happened as such.

Stretton refers constantly in his books to working out of the Darwin police station from Christmas Day onwards. He had first arrived at the Police Station late at night on Christmas Day, or in the early morning of Boxing Day – he is not specific in his autobiography. He does mention a non-specific number of bodies "in rows" in the station: "in a large room that had been pressed into service as a mortuary."[10] Police Sergeant Alex Carolan says "By midmorning there were five bodies in the corridor." Carolan then mentions Casuarina Police Station, about twelve kilometres away, being used as a temporary morgue – it would have made sense to concentrate this task into one building.[11] It seems the Casuarina post office was indeed used

---

7   Cunningham, Sophie. *Warning: the story of Cyclone Tracy.* Victoria: Text Publishing, 2014. (p. 67)

8   Slater, Dee. *Tracy's Fury.* Self-published: Warwick, Qld; 2018. (pp:71-72)

9   Department of Defence. *The Defence Force in the Relief of Darwin after Cyclone Tracy.* Australian Government Printing Service, 1980. (p.14)

10   Stretton, Alan. *The Furious Days.* Sydney: Collins, 1976. (p. 46)

11   *The Advertiser.* "Cyclone Tracy 40 Years On - Part 1: Impact & Survival – The Hardest Job Of All". 25 Nov 2014. https://www.adelaidenow.com.au/news/special-features/cyclone-tracy-impact-and-survival/news-story/69287b44fdc4ae1a6954f69 46086f6b0 Accessed June 2021.

as a temporary morgue: Sophie Cunningham interviewed one of the policemen who was stationed there who confirms it. [12]

The story of the Darwin Police Station being used extensively as a morgue isn't borne out by other accounts. Barbara Huddy, from the nearby suburb of Stuart Park, was at the police station that day "where she was a temporary guest", but although she describes the bad behaviour of some of the prisoners from Fannie Bay jail present, she says nothing of bodies.[13]

Former police officer then Sergeant Daryl Manzie, later to become a minister in the NT government, remembers well his time working at the Darwin police station post-Tracy. He says:

> I never saw any bodies as such from my arrival on day three of the post-cyclone period, although of course that doesn't mean there were not some present earlier. But if it had been the case that scores of bodies had been there it would have been the talk of the station.[14]

There was considerable effort put in by the police to do things properly. Commissioner Bill McLaren, recalls Daryl, later required every policeman who was in Darwin in the cyclone to write down what they did during and after Tracy's impact. Regarding those deceased "the coroner's constable Alex Carolan was taking a photo of them with details, and these were pinned up on a wall, for possible identification."[15]

The most likely reality of Slater's story is that yes, a few bodies were temporarily in Darwin Police Station, but they were soon transferred to Casuarina. As to there being 50, that would seem impossible, as a total of 66 fatalities is certain, as will be shown, and of those 21 were never recovered, being lost at sea, giving a maximum of 45 which could have arrived in storage.

The mass graves allegation

The next part of the fatalities controversy concerns allegations that many more than were being officially admitted had died, and furthermore, those fatality numbers were being concealed.

Stretton and the police commissioner, in their twice daily briefings to members of the Press:

> … were continually denying the rumour that the true number of dead was being concealed and that masses of dead were being buried in mass graves. I am not sure how these rumours started but they persisted for some time well after the end of the emergency. They were, of course, completely false.[16]

The wreckage was being continually dug through, and people were being found, but they were in many cases alive although injured. The *Canberra Times* reported on 31 December:

> One woman, 77, who was found alive under the rubble of her home 3½ days after the cyclone, was said to be in a very serious condition in Darwin Hospital yesterday. Arrangements were being made to fly her to Sydney.[17]

One of these stories seems to have emanated from the ill-starred entertainer Rolf Harris. He

12 Cunningham, Sophie. *Warning: the story of Cyclone Tracy.* Victoria: Text Publishing, 2014. (p. 65)

13 McKay, Gary. *Tracy: the storm that wiped out Darwin on Christmas day 1974.* NSW: Allen and Unwin, 2004. (pp: 171-172)

14 Manzie, Daryl, AM. Former police sergeant, NT Police. Interview with the author, 19 June 2023.

15 Manzie, Daryl, AM. Former police sergeant, NT Police. Interview with the author, 19 June 2023.

16 Stretton, Alan. *The Furious Days.* Sydney: Collins, 1976. (p. 92)

17 *The Canberra Times.* "ACT men search for bodies." 31 Dec 1974. (p. 1)

had made his way to Darwin in the days following the cyclone. Resident Beverley Wilson remembers his performance was staged at the amphitheatre at the Botanical Gardens.[18] At the time Harris was famous for his act of singing quirky or melodramatic songs, complete with a "wobble-board" and a didgeridoo, several of which, such as "Tie me kangaroo down sport", "Jake the Peg", and "Sun Arise" becoming international hits. He had regular TV appearances and was an accomplished painter who later in life went on to even more fame, at one stage painting Queen Elizabeth.[19] He made media statements about bodies on his return south following his performance in the stricken city:

> The Northern Territory Police Commissioner, Mr Bill McLaren, strongly denied today a statement by entertainer Rolf Harris in Sydney that the Cyclone Tracy death toll was much higher than the official figure.

> "I wish people like Harris wouldn't make sensational statements without checking things first", he said. Mr McLaren said the death toll attributable to Cyclone Tracy still stood at 49.

> Mr Harris had said that bodies were being found daily in the rubble of Darwin and that he gathered "the authorities are not willing to release exact figures for fear of creating a scare". He had said that while he was in Darwin salvage teams found the bodies of a woman and a child in the bath of a collapsed home. Mr McLaren said a search of all the houses in Darwin had been completed last night and no bodies adding to the official death toll had been recovered.

> He said naval authorities who made a further search of some areas covered by police had also failed to find any bodies.[20]

The Darwin concert, incidentally, was followed by a benefit event staged in Sydney. Performers included Olivia Newton-John, Barry Humphries, Barry Crocker, Dame Joan Hammond, Brian Cadd, Johnny Farnham, Colleen Hewett and Paul Hogan.[21]

Gary McKay, in *Tracy*, refers to a truck driver quoted in the *Sunday Observer* in a 1977 article. He said he was involved with the transport and burial of hundreds of bodies. McKay also says although he himself was contacted for his own research by the wife of a police officer involved in a search for supposed burial sites, nothing was found.[22]

Sophie Cunningham relates a story of a journalist who knew another journalist who she interviewed; the former had been "dining out on his cyclone experiences which included helping to personally load two hundred bodies on the back of a truck at Katherine – or to go to Katherine".[23] To load 200 bodies onto trucks would have taken a number of such vehicles, and a number of people to do it. Why they would then take the bodies 300 kilometres south is curious reasoning: if the authorities wished to bury these people there are enormous amounts of crown land around Darwin to utilise. No-one connected with such activity has come forward.

18   Wilson, Beverley. NT Government worker in 1974-5 for the organisation of the permit system. Interview September 2023, and emails 2024.

19   Harris's career crashed in March 2013, when he was one of twelve people arrested in the UK in Operation *Yewtree*, concerning historical allegations of sexual offences. He was tried and convicted of 12 offences of indecent assault, one of which was later overturned. He served three years in prison. He died in 2023.

20   *The Canberra Times*. "Police deny death toll higher". 11 Jan 1975. (p. 7)

21   *The Canberra Times*. "Seats for Darwin concert rushed." 31 Dec 1974. (p. 1)

22   McKay, Gary. *Tracy: the storm that wiped out Darwin on Christmas day 1974*. NSW: Allen and Unwin, 2004. (pp: 193-194)

23   Cunningham, Sophie. *Warning: the story of Cyclone Tracy*. Victoria: Text Publishing, 2014. (p. 70)

But the stories continued. Later, Dee Slater puts a figure on how many she says died in total: "inestimable thousands lost their lives to Cyclone Tracy. That is a fact!"[24]

Bill McLaren, the police commissioner, said in an interview later about the stories:

> Yeah, I'm quite aware of them. But I'm quite adamant in this and I was most insistent at the time – that every body, and I think this could be proved by checking cemetery records – every body had a separate grave, and every body had a proper funeral, according to whatever rights could be administered at the time. But each and every body had what respect could be given to it at the time. And most definitely, there were no bodies buried in mass graves.[25]

The rumours continue to the present day, as one website in 2011 attested:

> I had been told by a friend, a long-time resident of the Top End, that her brother had bulldozed into pits, and covered, lots of bodies, and the real death toll was hundreds. The old Darwin was a place people went to drop off the grid, and no-one knew how many aboriginals were camped around the town at the time.[26]

Rumours like this, the author can attest, make their way into public consciousness and then the media following massive disasters. They were prevalent in, coincidentally Darwin, following the first massive air raid of 19 February 1942. Hundreds, then thousands more than the 236 dying were alleged. One soldier said "We buried at least 300 bodies in one mass grave at Mindil Beach. From my estimation the losses were anywhere from six hundred to one thousand".[27]

The manager of the Commonwealth Bank wrote to the Governor of the Bank that: "I can assure you that between 600 and 1,000 lives were lost".[28] People visiting sites of conflict later embellished the stories. Queensland Senator Herbert Yeates visited Darwin towards the end of the war. He later wrote: "From the definite and reliable information I obtained whilst at Darwin … I am satisfied that about 1,200 people were killed but that includes merchant and other seamen, such as men from foreign ports."[29]

Later accounts of life in the Northern Territory perpetuate the stories. The popular writer Judy Nunn's novel *Territory* suggests in a foreword note regarding the war that "The true casualty figure is estimated to be in excess of 500." Roland Perry says in his history book *Centre Stage*, that as many as 1,100 died and were buried in mass graves at Mindil Beach.[30] The fictional 2008 movie *Australia* didn't help matters by portraying Japanese forces landing in Darwin, which they never did. The body count in the film was at least unresolved.[31]

24 Slater, Dee. *Tracy's Fury*. Self-published: Warwick, Qld; 2018. (p. 160)

25 Mclaren, Bill. Library & Archives NT, Northern Territory Archives Service, NTRS 226. Typed transcripts of oral history interviews with 'TS' prefix, TS 586.

26 *This Adventurous Age*. Website. https://thisadventurousage.com/tag/cyclone-tracy/ Accessed Jan 2022.

27 Rayner, Robert. *The Army and the Defence of Darwin Fortress*. NSW: Rudder Press, 1995. Private Alan R. Dunstone, 2/4 Pioneer Battalion AIE, statement p. 228.

28 Commonwealth Bank Manager to Governor of the Bank. Letter. "Evacuation of Darwin Branch." NSW. 24 March 1942. Alice Springs. (Copy in possession of the author, provided courtesy of Jack Mulholland, AA Gunner, Australian Army, August 2010.) A copy of this letter can be seen in the displays of the Darwin Military Museum.

29 Bradford, John. Report commissioned by the Darwin City Council. "Inquiry into enemy air raids on Darwin." War Cabinet Agendum No.116/1942. NAA Series No. A5954, Control Symbol No. 524/4, page 120. 2009.

30 *NT News*. http://www.ntnews.com.au/article/2012/02/17/289151_ntnews.html "Historian denies Bombing toll was higher." 17 February 2012.

31 *Australia – a Baz Lurhmann film*. http://www.australiamovie.net Movie released November 2008.

Bill and Polly Day seen two days after the cyclone with their damaged caravan in McMinn Street, Darwin. The city was home to a reasonable population of residents not living in regular housing and that has been used as an argument for why the death toll is allegedly under counted. (NT Library)

The media over the years often cannot resist the temptation to hint at higher numbers. A *Women's Day* feature in 2004 said:

> Officially, 65 people – 49 on land and 16 at sea – were killed during Cyclone Tracy's rage. But locals say anecdotal evidence points to many more who were never accounted for – including hippies who were camped on the city's beaches and itinerants who were using Darwin's remoteness as a place to hide.[32]

However, the reasoning behind such allegations must be challenged:

1. Just because a witness was there at the time does not mean they must be 100% credible. This is a very common fallacy: such people *must* be believed. But this writer has seen people swear to supposed facts: a German-swastika marked aircraft in a Darwin air raid, for example, or a second submarine periscope in an attack – both incidents where records later verified these were not possible. Thus myths are born.

2. People can be deceived; they can be traumatised and shocked, and they shut their eyes, imagining what is in front of them. Note the example of the Commonwealth Bank Manager above, whose bank was indeed in the centre of town close to sites of great damage – the Darwin Post Office for example, destroyed with severe loss of life. Yet in the very same letter he charts his own progress, and that of his staff, over the hours following the first two raids, during which attacks he ventured no further from the bank than the bombed post office about 45 seconds walk away. His afternoon and the next morning were full of meetings and decision making, during which the bank was closed and its

---

32   *Women's Day.* "Cyclone Tracy 30 Years On." 27 December 2004 (p. 27)

cash together with commercial papers was loaded on trucks. At 0100 the day following the raids – that is, some 24 hours after the second raid had ended – he journeyed south down the main highway out of town to Adelaide River and beyond, never to return. How could he possibly see the more than 200 confirmed people killed, who died over several areas, many kilometres apart all around Darwin?

3. What possible reason could "the government" have for a) carrying out secret Tracy burials, and b) concealing them? Although mass burials occur in wartime for hygiene reasons – there were such internments at Darwin's Mindil Beach in 1942, although the bodies were later removed to Adelaide River War Cemetery – what motive could there be for concealing them after the cyclone? To cover up government incompetence? The NT semi-government of the time was not guilty of anything involving neglect insofar as can be seen, and indeed the federal government units, such as the ADF, performed with distinction during and after the storm.

4. If thousands of people died as is claimed in the Darwin air raids, where are the many more thousands of relatives who would have followed up the losses after the war? The same logic applies to Cyclone Tracy. As an NT police officer, Daryl Manzie, there on the scene, pointed out later: "there's never been anyone who put their hand up and said my cousin, or my father, or brother is missing". To allege that they were "aboriginals" and somehow their families did not pursue details of their deaths is patronising.

5. Where are the scores if not hundreds of "government personnel" who carried out such disgraceful actions? By inference, they would have to have been people such as ADF members with access to heavy equipment for transportation and for digging mass graves. Stretton would have had to have known of such activities and indeed ordered them. Why have these people kept silent for 50 years? The answer is they haven't – because there were no such people engaged in such activities.

6. And lastly, if such graves existed where are they? Darwin has grown three-fold in 50 years: its population along with its satellite city of Palmerston is now around 150,000, and the attendant land use similarly expanded. Nothing to this author's knowledge, and that includes 30 years of military history analysis of actions around the area, has uncovered such sites.

As a measure of their conscientious handling of such a sensitive matter, the NT police kept the body of a crewman from the sunken *Flood Bird* until 1989 when he was unidentified:

> Cyclone Tracy's last unidentified victim has been named, thanks to a process used to identify victims of South Australia's Truro murders. Northern Territory Police have said that George Roewer died on the trawler *Flood Bird* when it sank in Darwin Harbour on Christmas Day 1974. Using craniofacial superimposition, they matched a skull with photographs of the dead man. Mr Roewer's next of kin have been informed.[33]

Sergeant Trevor Wauchope, the officer in charge of the Northern Territory Missing Persons Department, pointed out at the time if there had still been people missing from Tracy, their relatives would have likely contacted the police when the Tracy mystery was featured in the national papers.[34] The NT Police advise that there are no such current inquiries.

---

33 *Sydney Morning Herald*. "Tracy's final victim." 3 March 1990. https://www.smh.com.au/opinion/in-the-herald-march-3-1990-20160216-gmvg0v.html Accessed June 2021.

34 Lewis, Tom. *Wrecks in Darwin Waters*. Sydney: Turton and Armstrong, 1990. (p. 74)

The true fatality count

There was no one definitive federal government sponsored report on Cyclone Tracy; and no royal commission or anything like it. Rather a myriad of smaller reports were written, many of them concerned with their organisation's chief reason-for-being; e.g. the Bureau of Meteorology's report is most concerned with climate and the weather.

Those reports accounting for the fatalities are given in ascending order below.

Number of fatalities	Explanatory Text	Source
49	"The actual number of deaths due directly to Cyclone Tracy in Darwin and the surrounding area will probably never be known; it could possibly be considerably more than 49. Because Darwin was the sort of place where many people from all parts of Australia came to stay and lost their identities, for a while at least, no accurate estimate of the number missing and dead may ever be possible."[35]	Commonwealth of Australia. *The Experience of Cyclone Tracy*. Canberra: Australian Government Publishing Service, 1981. (p. 125)
49	This two-page document has a typed "McLaren, Commissioner of Police NT" at its conclusion, with a reference number NORLAW AA 85381. Unfortunately, it has several names scored out.	National Archives of Australia. "Updated list of deceased persons who came to their deaths during cyclone [Tracy] at Darwin on 25/12/74 identified / not identified." Item number 7068518.
61	"Tracy's official death toll reached 45 in Darwin with another 16 deaths at sea"[36]	Risk Management Solutions. "Cyclone Tracy 30-Year Retrospective". 2005. https://forms2.rms.com/rs/729-DJX-565/images/scs_cyclone_tracy_30_retrospective.pdf
62	The Darwin City Council plaque "lists 49 names of those who died in the cyclone and the 13 who died at sea".[37] Plaque unveiled by Queen Elizabeth II on 26 March 1977 (See note below re a revision to the plaque)	Monument Australia. "Cyclone Tracy Memorial". https://monumentaustralia.org.au/themes/disaster/storms/display/80124-cyclone-tracy-memorial
65	"Tracy struck Darwin early on Christmas morning 1974, causing the deaths of 49 people and damage estimated at hundreds of millions of dollars. Another 16 people were posted as missing at sea."	Australian Government. Department of Science. Bureau of Meteorology. *Report on Cyclone Tracy 1974*. Australian Government Publishing Service, 1977. (p. 1)
65	"Navy records the fatalities as 49 people ashore and a further 16 at sea."[38]	Mitchell, Brett. Royal Australian Navy. "Disaster Relief - Cyclone Tracy and Tasman Bridge". https://www.navy.gov.au/history/feature-histories/disaster-relief-cyclone-tracy-and-tasman-bridge
65	"The death toll was 49, with a further 16 listed as missing at sea."	Murphy, Kevin. *Big Blow up North* (A History of Tropical cyclones in Australia's Northern Territory). Darwin: NT Government (University Planning Authority), 1984. (p. 58)

---

35 Commonwealth of Australia. *The Experience of Cyclone Tracy*. Canberra: Australian Government Publishing Service, 1981. (p. 125)

36 Risk Management Solutions. "Cyclone Tracy 30-Year Retrospective". 2005. https://forms2.rms.com/rs/729-DJX-565/images/scs_cyclone_tracy_30_retrospective.pdf

37 Monument Australia. "Cyclone Tracy Memorial". https://monumentaustralia.org.au/themes/disaster/storms/display/80124-cyclone-tracy-memorial

38 Mitchell, Brett. Royal Australian Navy. "Disaster Relief - Cyclone Tracy and Tasman Bridge". https://www.navy.gov.au/history/feature-histories/disaster-relief-cyclone-tracy-and-tasman-bridge

Number of fatalities	Explanatory Text	Source
65	"Officially, 65 people - 49 on land and 16 at sea - were killed during Cyclone Tracy's rage. But locals say anecdotal evidence points to many more who were never accounted for - including hippies who were camped on the city's beaches and itinerants who were using Darwin's remoteness as a place to hide."	*Women's Day.* "Cyclone Tracy 30 Years On." 27 December 2004 (p. 27)
65	"The official death toll was 65"	National Archives of Australia Cyclone Tracy – excerpt from ABC documentary *The Darwin Story*
66	"The death toll from Cyclone Tracy was officially 50, with another 16 lives lost at sea."	McKay, Gary. *Tracy: the storm that wiped out Darwin on Christmas day 1974.* NSW: Allen and Unwin, 2004. (p. 174)
66	"Cyclone Tracy killed at least 66 people"[39]	Library and Archives Nt. "Explore_NT_History_Cyclone_Tracy_20180522". https://lant.nt.gov.au/explore-nt-history/cyclone-tracy 22 May 2018. (p. 5)
66	Darwin City Council plaque 2. At some stage the initial 1977 plaque was revised. The name list was revised, with a section below of 13 additional names added "and those who were lost at sea that night". Curiously, several names of those who were verified on ships are included in the upper list. The total of names is 66.	The original plaque with HM QEII can be seen with a photo of Her Majesty here: https://static.wixstatic.com/media/9dd312_502069542ca64b9d8274e2f0c19b3876~mv2.jpg/v1/fill/w_1000,h_694,al_c,q_85,usm_0.66_1.00_0.01/9dd312_502069542ca64b-9d8274e2f0c19b3876~mv2.jpg The second plaque looks to be a thorough revision of the first
66	Records have identified 66 names of individuals who perished as a result of the cyclone (53 on land and 13 at sea).[40]	National Archives of Australia. Fact Sheet 176. "Cyclone Tracy, Darwin". https://www.naa.gov.au/sites/default/files/2020-05/fs-176-cyclone-tracy-darwin.pdf
71	"Tropical Cyclone Tracy is arguably the most significant tropical cyclone in Australia›s history accounting for 71 lives …"	Australian Government. Department of Science. Bureau of Meteorology. "Severe Tropical Cyclone Tracy." http://www.bom.gov.au/cyclone/history/tracy.shtml Undated one page article

Notably above is the jump from 65 lives to 71 fatalities recorded by the same Department in 1977.

It may be that the reporting added five names but they were already in the list, so 66 + 5 makes 71. However, there were six names in the Coroner's Report, five from the *Booya* and one from the *Darwin Princess*.

Number of fatalities	Explanatory Text	Source
71	Wikipedia: "Tracy killed 71 people…" Sources given are: The ABC 2005 report on the Coroner's findings on the loss of life in the vessels *Booya* and *Darwin Princess*, which has been recently discovered, and The Attorney-General's Department Disasters Database, which says: "The cyclone resulted in the death of 71 people, which included 22 who were lost at sea."	Australian Broadcasting Corporation. AM (Radio program) "NT coroner hands down finding on Cyclone Tracy deaths." https://web.archive.org/web/20050405142927/http://www.abc.net.au/am/content/2005/s1326359.htm Original broadcast 18 March, 2005. Attorney-General's Department Disasters Database: http://www.disasters.ema.gov.au/Browse%20Details/DisasterEventDetails.aspx?DisasterEventID=2174
71	"Cyclone Tracy, which hit Darwin in the small hours of Christmas Day 1974, killed 71 people and devastated 80 per cent of the city."	National Museum Australia. "DEFINING MOMENTS Cyclone Tracy." https://www.nma.gov.au/defining-moments/resources/cyclone-tracy#:~:text=Cyclone%20Tracy%2C%20which%20hit%20Darwin,per%20cent%20of%20the%20city

---

39 Library and Archives NT. "Explore_NT_History_Cyclone_Tracy_20180522". https://lant.nt.gov.au/explore-nt-history/cyclone-tracy 22 May 2018. (p. 5)

40 National Archives of Australia. Fact Sheet 176. "Cyclone Tracy, Darwin". https://www.naa.gov.au/sites/default/files/2020-05/fs-176-cyclone-tracy-darwin.pdf

Number of fatalities	Explanatory Text	Source
71	"Cyclone Tracy claimed the lives of at least 71 people." However, the site only lists 64, omitting any mention of Arthur Lim, or Arthur Fong Lim.	*Cyclone Tracy. Remembering Cyclone Tracy, fifty years later* (Website) "Remembering those who died." https://cyclonetracy.au/remembering-those-who-died/
71	Australian Parliament. Federation Chamber Private Members' Business - Cyclone Tracy. Speech on 40th anniversary. 1 December 2014.	https://www.aph.gov.au/Parliamentary_Business/Hansard/Hansard_Display?bid=chamber/hansardr/2d891fab-c2b5-41b4-967f-0b37fdb6fe7c/&sid=0223
	Notably this pair of speeches from the two NT members of the House of Representatives in 2014 mentioned the number of fatalities was 71. Leave was given to read the list of names into Hansard. This was done but the published list is only 66 names. It appears that the list used was that published by the *NT News*.	
71	Sophie Cunningham's book says "around seventy-one people died," [41 and]: "The official number of deaths at sea now sits at twenty-two." [42] "...the current official missing list is a hundred and sixty people."[43]	Cunningham, Sophie. *Warning: the story of Cyclone Tracy.* Victoria: Text Publishing, 2014. (pages 6, 57, and 79)
71	The number is now used in a variety of government and business websites, including but not limited to those listed to the right here:	National Museum of Australia The Australian Institute for Disaster Resilience Bureau of Meteorology

People involved in Tracy do generally have an attitude there was a cover up. Some years later a poll was taken which established that there was indeed considerable public suspicion that there had been a governmental concealment regarding the fatalities number. The table is reproduced verbatim[44] below:

Belief in the accuracy of the published death toll	Non-returned evacuees (number surveyed = 219)	Returned evacuees (number surveyed = 107)	Stayers (number surveyed = 90)
Yes, accurate	24%	15%	36%
Not accurate	64%	70%	58%
No answer, don't know	12%	15%	7%

So those who had been evacuated from the NT; and those who later returned had the highest suspicion that more people had died than the government were saying. This suspicion was lowest amongst those who had remained in the Territory.

---

41 Cunningham, Sophie. *Warning: the story of Cyclone Tracy.* Victoria: Text Publishing, 2014. (p. 6)

42 Cunningham, Sophie. *Warning: the story of Cyclone Tracy.* Victoria: Text Publishing, 2014. (p. 57)

43 Cunningham, Sophie. *Warning: the story* of Cyclone Tracy. Victoria: Text Publishing, 2014. (p. 79)

44 Commonwealth of Australia. *The Experience of Cyclone Tracy.* Canberra: Australian Government Publishing Service, 1981. (p. 129) Table is given verbatim; the RH column in the original does add up to 101%

The Fatality List

The list shown below is the result of considerable research. Its columns reflect the most prominent records and are retained to add for the reader an understanding of the analysis used to arrive at the final total of 66.

No.	Name from Hansard Plus additional information on spelling variations etc	Further details Information from Cyclone Tracy project[45] unless otherwise noted	Recorded in Hansard?	Recorded in National Archives 7068518	Recorded on DCC Plaque Mk 1 or 2 [46]
1	Andrew, Peter		Hansard	No	Yes 2
2	Bell, Malini Palathil		Hansard	NAA 7068518	Yes 1
3	Bloomfield, Dorothy		Hansard	NAA 7068518	Yes 1
4	Bonner, Cecil Henry	Possibly English, aged 66, born in South Hampton in 1918, an ex-British soldier	Hansard	NAA 7068518	Yes 1
5	Brown, Geraldine Elizabeth[47]	Daughter of Kay Brown and sister to Stephanie and Christine. 8 years old.	Hansard	NAA 7068518	Yes 1
6	Bruhn, Andrew Mark	Aged three	Hansard	NAA 7068518	Yes 1
7	Bolger, Wotsanbuk (Bungun Major)		Hansard	NAA 7068518	Yes 1
8	Burgess, Dean William	Aged two years, seven months	Hansard	NAA 7068518	Yes 1
9	Butler, Louisa Fanny		Hansard	NAA 7068518	Yes 1
10	Catton, Leslie Kevin [48] [49]	At sea – HMAS *Arrow*	Hansard	NAA 7068518	Yes 1
11	Chaney, Paul Mark	Aged 6 months	Hansard	NAA 7068518	Yes 1
12	Clarke, Rose Susanna		Hansard	NAA 7068518	Yes 1
13	Clough, Eileen Patricia May	Aged 33, mother of three, and wife to husband Colin	Hansard	No	Yes 1
14	Curtain, Raymond John	At sea – captain and co-owner of *Darwin Princess* [50]	Hansard	No	Yes 2
15	Daffey, Peter Brian[51]		Hansard	NAA 7068518	Yes 1
16	Daniel, Avis Winifred	At sea – *Mandorah Queen* [52] Nursing sister at Darwin hospital, aged 21	Hansard	No	Yes 2
17	Dearden, Graham William	At sea – *Booya* [53]	Hansard	No	Yes 2

---

45 https://cyclonetracy.au/category/remembering-those-who-died/ Information as to April 2024. The project's aim was to gather and post information, so more may be possibly seen at the website.

46 If noted as being on plaque 2 that means the name was not present on plaque 1.

47 Cunningham, Sophie. War*ning: the story of Cyclone Tracy.* Victoria: Text Publishing, 2014. (p. 66)

48 In Hansard given as "Calton"

49 Oddly, Catton and Rennie who died on in the sinking of HMAS *Arrow*, are not in the "died at sea" section.

50 Lewis, Tom. *Wrecks in Darwin* Waters. Sydney: Turton and Armstrong, 1990. (p. 74)

51 Cunningham, Sophie. *Warning: the story of Cyclone* Tracy. Victoria: Text Publishing, 2014. (p. 67)

52 Lewis, Tom. *Wrecks in Darwin Waters.* Sydney: Turton and Armstrong, 1990. (p. 71)

53 Lewis, Tom. *Wrecks in Darwin* Waters. Sydney: Turton and Armstrong, 1990. (p. 75)

No.	Name from Hansard Plus additional information on spelling variations etc	Further details Information from Cyclone Tracy project[45] unless otherwise noted	Recorded in Hansard?	Recorded in National Archives 7068518	Recorded on DCC Plaque Mk 1 or 2 [46]
18	Dewar, Peter James		Hansard	NAA 7068518	Yes 1
19	Dibua, Charles Joseph		Hansard	NAA 7068518	Yes 1
20	Fealy, David Grant	At sea – *Flood Bird* [54]	Hansard	No	Yes 2
21	Fenton, Gary Roger [55]		Hansard	NAA 7068518	Yes 1
See Lim, Arthur	Fong Lim, Arthur[56]		Hansard	NAA 7068518	No
22	Grant, Michael John	At sea – *Mandorah Queen* [57]	Hansard	No	Yes 2
23	Hampton, Raymond		Hansard	NAA 7068518	Yes 1
24	Hanson, Thomas William		Hansard	NAA 7068518	Yes 1
25	Harder, Lisa Karoline		Hansard	NAA 7068518	Yes 1
26	Hoff, Donald James Duncan	At sea – *Dieman* [58] Born 1938, "late of New Zealand"	Hansard	NAA 7068518	Yes 1
27	Holten, Dennis Keith	At sea – *Flood Bird* [59] Aged 23	Hansard	NAA 7068518	Yes 1
28	James, Rueben Paul	22 days	Hansard	NAA 7068518	Yes 1
29	James, Richard Roland	24 years	Hansard	NAA 7068518	Yes 1
30	Knox, Catherine Michelle	Michael and Elva Knox and their two children Stephen (3 years) and Catherine (5 months) were living in Adcock Crescent, Nakara, at the time in the northern suburbs, one of the worst hit areas.	Hansard	No	Yes 1
31	Knox, Elva Dawn		Hansard	No	Yes 1
32	Knox, Michael John		Hansard	NAA 7068518	Yes 1
33	Knox, Stephen James		Hansard	No	Yes 1
34	Lang, John Dudley (Jack)		Hansard	NAA 7068518	Yes 1
35	Lai, On'ing Owen		Hansard	NAA 7068518	Yes 1
36	Lim, Arthur [60] [61]		Hansard		Yes 1
37	Marshall, Ronald Charles	Lost at sea – *Charles Todd* [62]	Hansard	No	Yes 2

54 Lewis, Tom. *Wrecks in Darwin Waters.* Sydney: Turton and Armstrong, 1990. (p. 69)

55 Referred to as "Harry" in the NAA document

56 Niece Mayor Katrina Fong Lim confirms his death by accident in the cyclone. https://www.sbs.com.au/news/article/darwin-marks-the-day-tracy-blew-christmas-away/i12tf1bsg He is confusingly recorded as Lim, Fong, Arthur on the DCC plaque and in Hansard/NT News as "Lim, Arthur".

57 Lewis, Tom. *Wrecks in Darwin Waters.* Sydney: Turton and Armstrong, 1990. (p. 71)

58 Lewis, Tom. *Wrecks in Darwin Waters.* Sydney: Turton and Armstrong, 1990. (p. 70)

59 Lewis, Tom. *Wrecks in Darwin Waters.* Sydney: Turton and Armstrong, 1990. (p. 69)

60 Recorded twice – as Lim Arthur, and as – correctly Fong Lim, Arthur.

61 Niece Mayor Katrina Fong Lim confirms his death by accident in the cyclone. https://www.sbs.com.au/news/article/darwin-marks-the-day-tracy-blew-christmas-away/i12tf1bsg He is confusingly recorded as Lim, Fong, Arthur on the DCC plaque and in Hansard/NT News as "Lim, Arthur". She advises that the use of "Lim, Arthur" is the spelling preferred by the family.

62 Lewis, Tom. *Wrecks in Darwin Waters.* Sydney: Turton and Armstrong, 1990. (p. 73)

No.	Name from Hansard Plus additional information on spelling variations etc	Further details Information from Cyclone Tracy project[45] unless otherwise noted	Recorded in Hansard?	Recorded in National Archives 7068518	Recorded on DCC Plaque Mk 1 or 2 [46]
38	McNab, Eric Arthur[63]		Hansard	NAA 7068518	Yes 1
39	Macklin, Paul Alaister [64]	Senior specialist and anaesthetist at Darwin Hospital	Hansard	NAA 7068518	Yes 1
40	Muir, William Daniel	Born in 1916, enlisted into the Army on the first day of WWII, 4 Sep 1939, and was discharged 10 Oct 1945 as a corporal, having served in the infantry.[65]	Hansard	NAA 7068518	Yes 1
41	Odawara, Shigemori	At sea – *Flood Bird* [66]	Hansard	NAA 7068518	Yes 2
42	Parker, George Valentine		Hansard	NAA 7068518	Yes 1
43	Portman, Suzanne Mary		Hansard	NAA 7068518	Yes 1
44	Rennie, Ian Robert	At sea – HMAS *Arrow* [67]	Hansard	NAA 7068518	Yes 1
45	Roewer, George	At sea – *Flood Bird* [68]	Hansard	No	Yes 2
46	Smaniotto, Giovanni		Hansard	No	Yes 1
47	Stephenson, Cherry Leonie Rose	Aged 22, wife of Geoff Stephenson, a sailor on HMAS *Arrow*	Hansard	NAA 7068518	Yes 1
48	Stephenson, Kylie Jane		Hansard	NAA 7068518	Yes 1
49	Swann, Robert Gordon	At sea – *Flood Bird* [69]	Hansard	No	Yes 2
50	Taylor, John Thomas		Hansard	No	Yes 1
51	Thompson, Gerald Frederick	At sea – *Booya* [70]	Hansard	No	Yes 2
52	Thompson, Richard		Hansard	NAA 7068518	Yes 1
53	Turner, John	Aged 69 years	Hansard	NAA 7068518	Yes 1
54	Vincent, Ruth Nazmeena Adrione [71]	At sea – *Booya* [72]	Hansard	No	Yes 2
55	Wade, Robert Norman	At sea – *Frigate Bird* [73]	Hansard	No	Yes 2
56	Wenck, Shirley		Hansard	NAA 7068518	Yes 1

63 Second McNabb on the DCC plaque?

64 Recorded as Macklin on the DCC plaques 1 and 2

65 Australian Government. Department of Veterans Affairs. Muir, William Daniel. WWII Post-nominal roll. https://nominal-rolls.dva.gov.au/veteran?id=1205325&c=WW2#R

66 Lewis, Tom. *Wrecks in Darwin Waters*. Sydney: Turton and Armstrong, 1990. (p. 69)

67 Oddly, Catton and Rennie who died on in the sinking of HMAS *Arrow*, are not in the "died at sea" section

68 Lewis, Tom. *Wrecks in Darwin Waters*. Sydney: Turton and Armstrong, 1990. (p. 69)

69 Lewis, Tom. *Wrecks in Darwin Waters*. Sydney: Turton and Armstrong, 1990. (p. 69)

70 Lewis, Tom. *Wrecks in Darwin* Waters. Sydney: Turton and Armstrong, 1990. (p. 75)

71 Ms Vincent's name often includes her third name as "Adrienne". The coroner's report, which had access to her birth certificate, uses "Adrione."

72 Lewis, Tom. *Wrecks in Darwin* Waters. Sydney: Turton and Armstrong, 1990. (p. 75)

73 Lewis, Tom. *Wrecks in Darwin Waters*. Sydney: Turton and Armstrong, 1990. (p. 69)

No.	Name from Hansard Plus additional information on spelling variations etc	Further details Information from Cyclone Tracy project[45] unless otherwise noted	Recorded in Hansard?	Recorded in National Archives 7068518	Recorded on DCC Plaque Mk 1 or 2 [46]
57	Westerman, Gregory John	At sea – *Booya* [74]	Hansard	No	Yes 2
58	Westwood, Terrence Carew	At sea – *Booya* [75]	Hansard	No	Yes 2
59	Wheatley, Kenneth James Scott	Son of Cherry Stephenson [76]	Hansard	NAA 7068518	Yes 1
60	Williams, Kerry Lynda		Hansard	NAA 7068518	Yes 1
61	Williams, Molly		Hansard	NAA 7068518	Yes 1
62	Wolfe, Siegfried Karl Otto[77]	At sea – *Mandorah Queen* [78]	Hansard	No	Yes 2
63	Wood, Jennifer Anne		Hansard	NAA 7068518	Yes 1
64	Woodyatt, William [79]	At sea – *Frigate Bird* [80]	Hansard	No	Yes 2
65	Yoshida, Koji	At sea – *Dieman* [81]	Hansard	NAA 7068518	Yes 1
66	Yu, Frederick Yee Char	Aged 36 years	Hansard	NAA 7068518	Yes 1
	Lost on land total = 45 Lost at sea total = 21				

Notes

- Total Hansard names = 66 but 71 was referred to by both MLAs.

- Total National Archives Item number 7068518 names contains 49 names however only 44 are legible; the others appear to have been scored out.

- Total Darwin City Council plaque #1 names = 49

  - Presumably it used the 44 names. However. it subtracted two: Odawara, Shigemori and Turner, John. It has 42 of the NAA document. It added six:

    - Clough, Eileen
    - Knox, Elva
    - Knox, Stephen
    - Knox, Catherine
    - Smaniotto, Giovanni
    - Taylor, John Thomas

---

74 Lewis, Tom. *Wrecks in Darwin* Waters. Sydney: Turton and Armstrong, 1990. (p. 75)

75 Lewis, Tom. *Wrecks in Darwin* Waters. Sydney: Turton and Armstrong, 1990. (p. 75)

76 Cunningham, Sophie. *Warning: the story of Cyclone* Tracy. Victoria: Text Publishing, 2014. (p. 67)

77 Surname was given in Hansard as "Waffle". In the original NT News document it is "Wolfie". On the second DCC plaque it is "Wolfe".

78 Lewis, Tom. *Wrecks in Darwin Waters*. Sydney: Turton and Armstrong, 1990. (p. 71)

79 Surname sighted on DCC plaque as "Woodyatt"??? Also in Remembering Tracy site as "Woodyatt" – as it is in NT News list. Also on DCC Plaque 2.

80 Lewis, Tom. *Wrecks in Darwin Waters*. Sydney: Turton and Armstrong, 1990. (p. 69)

81 Lewis, Tom. *Wrecks in Darwin Waters*. Sydney: Turton and Armstrong, 1990. (p. 70)

- However, that only makes 48. The confusion is that "Fong Lim, Arthur" appears in the NAA document under that name, but as "Lim, Arthur" in the Hansard List.

- The Hansard list, as advised by the Table Office, Department of the House of Representatives, was taken from the 40th edition of the *NT News*, and not compiled by the office's staff.[82]

- Plaque #2 has 66 names. Advice from the Museum and Art Gallery of the NT says that the second version of the plaque was unveiled in 2014, although it retained the wording that it had been unveiled by Queen Elizabeth.

In conclusion, there is strong evidence that we now know the true number of people who died in Cyclone Tracy, and we know their names. The research above has been agreed to by the Northern Territory Police, Fire and Emergency Services.[83] There were 66 fatalities, of which 45 died on the land and 21 were lost at sea. May they Rest in Peace.

---

82   Australian Parliament. House of Representatives. Table Office, Department of the House of Representatives. Emails February 2024.

83   Email from Information Officer Ms Genevieve Reed, Northern Territory Police, Fire and Emergency Services. 29 Feb 2024.

CONCLUSION

CYCLONE TRACY AND THE ARMED FORCES

The armed forces responded magnificently to the disaster that was Cyclone Tracy. Rallying from around Australia, at the start of the annual holiday season, and often responding without orders, they reported to their place of duty. Taking the initiative with a thought of "what needs to be done", they began rescue and relief operations that started immediately and lasted for months.

Major General Stretton was the man for the hour. He was the right choice: his army training and experience made him capable of weighing up a situation, analysing it at speed, and making a decision. He may have upset some people, but it is extremely doubtful the massive destruction to a city and the resultant total breakdown of essential services could have been temporarily rectified by those who were on the ground already. Remember, their personal lives – their families and living arrangements – had been shattered. Bringing someone in from the outside worked.

The Navy was hardest hit – its members saw loss of life in their own uniformed numbers, and also in a service family. They had seen the ultimate naval disaster – the loss of a ship – happen in their own harbour. Yet they immediately rallied round, and organised by most capable leaders, began work immediately to help and to salvage. The enormous force steaming steadily up the east coast to the rescue was the right plan delivered the right way.

The Air Force was on the scene immediately and will be associated forever with the quick evacuation of those civilians who needed to be moved south. The RAAF steadily handled a variety of challenges, from aerial mapping to aerial spraying to the freighting in of essentials. They kept the air effort going in a non-stop fashion for weeks.

The Army, although the smallest service in town at the time, eventually became the biggest. In sheer tonnes of material moved, the Army proved itself the strongest service. What's more, they kept at it for months.

Those in other uniforms proved themselves too. The police – both near and from far; the emergency services, Red Cross and Salvation Army were indispensable.

All of this together kept Darwin going. Without those in uniform the northern capital likely would have resulted in a consideration, weeks or months later, regarding its removal rather than its repair.

Some aspects remain unaddressed, however. There has not been sufficient reward. The bestowing of the National Emergency Medal on those involved should begin now and keep going until all uniformed people involved have been recognised. And a fully accurate memorial to all of Tracy's victims should be established on Darwin's Esplanade.[1]

---

1 In 2024 the federal Albanese Government announced a grant of $600,000 to fund two memorials - at the time of going to press these had not been finalised.

APPENDIX

THE CURIOUS CASE OF MAJOR GENERAL STRETTON VERSUS THE ARMY

One of the strangest aspects of the Army side of the Cyclone Tracy story is the animosity between Major General Stretton and his own former service. The overall commander of the disaster situation was critical of the Army from the start. It seemed this was both in his own judgement and also overt: with direct criticism coming from him verbally. It was later backed up with written castigation in his memoir.

Early on in Major General Stretton's term of command, it seems he singled the Army out for bitter comment, perceiving they were not pulling their weight. He had arrived on the evening of Christmas Day, and over the next two days he became embroiled in a clash with several senior Army members. Indeed, one author, publishing a book of Australian photographs almost a decade later, condemned all of the armed forces with a throwaway comment:

> … the armed forces behaved disgracefully, failing to assist rescue operations with their men, vehicles and rations.

However, there is no evidence cited to back this up.

The Army situation seems to have been set, at least in Stretton's mind, from his first arrival. He flew in on the night of Christmas Day, arriving "at 10.20pm precisely". He toured some parts of the damaged area; had some hours of rest from 0330, and then chaired a meeting the next morning, at which "Surprisingly, the Army was not represented." From then on, he lists the service's failures in his eyes:

> The Army in Darwin had shown little initiative and that the only action taken outside their barracks seemed to be the installation of the local radio network which had just been established at my direction.

> … the Army commander in Darwin accepted my authority with reluctance.

> The Army Commander had not appeared at the 9am conference so I contacted him by radio and directed him to attend the 5pm conference. He said he had not attended the 9am conference because he did not receive a specific invitation!

One interesting aspect of Stretton's leadership was he did not immediately call upon his own service. While this was understandable in the case of the Navy – they were organised and already responding, and the RAAF were very quickly engaged in evacuations one way, and supply runs inwards the other way, the Army were not tasked, with the exception of getting a signals officer to establish radio contact the night he arrived.

The services in Darwin consisted of 160 Army, 409 Navy and 670 Air Force personnel. Given Darwin's population of 46,656 on 30 January 1974, this shows the number of people who worked for the services was 1,239, or some 2.6% of the total. Dependents numbered for the Navy 287 and the RAAF 913. No report has been located which shows the number of Army dependents, but if it was 1:1 this would mean 2,599 people were dependent on the ADF, giving an approximate number of some 5.6% of Darwin residents being in or connected to the armed forces.

On 30 December Stretton levelled his most serious accusation against the Army:

> I then decided to visit the Army at Larrakeyah Barracks. This was the first time that I had the opportunity to do this. I called on the OC and asked him to accompany me around the barracks. Larrakeyah Barracks did not appear to be as extensively damaged as I thought. The mess and administrative buildings had stood up fairly well – the main damage seemed to have been to the married quarters. There was an air of unreality within the area of the barracks. The workshops were still intact, but the only soldier present was packing a trailer of personal belongings. He said the rest of the workshop staff had "been given a stand down day". It seemed absurd to find a fully equipped workshop completely ineffective in the middle of an emergency because soldiers had been given the day off.
>
> This was not all. Parked in the unit car park were a number of idle vehicles and urgently needed engineer plant. These vehicles were additional to the transport platoon that I had previously requisitioned. As we progressed around the barracks, further things came to light. I found 27,000 eatable rations in damaged containers. Despite the initial shortage of food after Cyclone Tracy, enquiries revealed that no effort had been made to distribute these rations to the homeless and hungry outside the boundaries of the barracks. It was obvious that the Army in Darwin, who had a strength of 153 on Christmas Day, had taken practically no initiative to help a community in distress outside the assistance given to their own families. On that morning, I came close to being ashamed of the service I loved so much.

Stretton also criticised an Army lieutenant colonel who had just been posted in for not coming forward to volunteer his services – the previous officer was still in the post.

Recounting the events of 28 December, Stretton wrote:

> Up to this point of time, the Army had done little outside the boundary of their barracks. They had shown little initiative of their own and it had been necessary for me to give direct orders about the provision of radio sets, the use of their vehicles (which had to be requisitioned) and assistance to the police in guarding vital points. It seemed as if Operation *Clean Up* [the collection and disposal of hazardous materials] would provide an opportunity for them to at last make a major contribution to the relief operations … I decided to put the senior Army officer in charge of the operation.

Connair's John Myers was also particularly critical of the lack of response in the short term from army personnel based at Larrakeyah Barracks. A former army reservist, Myers headed for the barracks on Christmas morning in an effort to get some help, only to find the gates locked. When he called to an army man to suggest those outside could do with some help, he was told the army had problems of its own.

Another complaint from Stretton involved the use of some Army trucks:

> I received a surprise visit from a major in uniform who called at the Darwin Police Station. He was in charge of the Supplies and Transport section at Larrakeyah Barracks. He had come to my headquarters at the police station to get authority for his platoon of vehicles to draw petrol. I could not believe that the Army had a full platoon of some sixteen 2.5-ton

vehicles which for the past two days had been engaged on camp duties when they were so urgently needed in the town to help with the distribution of food and relief supplies to the population of Darwin. The vehicles had only come to light because of the necessity for them to get authority to draw petrol.

Frank [Thorogood] skilfully summed up the situation when he heard the request and told the Major to see me if he wanted fuel. The Major's face was a study as I informed him I was requisitioning his vehicles which would immediately report to the Food Committee and work under the direction of the Chairman. I allowed two vehicles to be retained by the Army for their own needs. The Food Committee could not believe its good fortune when a convoy of Army trucks arrived to bring relief to the motley collection of damaged vehicles which were used for the distribution of food. I resolved that as soon as I could find a spare moment, I would make an inspection of Larrakeyah Barracks to see what other resources I could find that were being used solely for Army purposes.

Then again, the later official report takes a different line:

Of 20 x 2.5-ton vehicles, only two were serviceable. Two were unrepairable. As they became available, they were committed to "clearing roads within and to Larrakeyah Barracks and moving evacuees to the RAAF Base for air evacuation."

And

During the evacuation, water became of critical importance at staging and concentration areas such as Darwin airport. At midday on 28 December, NDO asked for jerrycans filled with water to be sent urgently to the city. A total of 585 cans, mainly from Townsville, were on their way by air that afternoon.

Stretton countermanded a decision by the Army to move in a "Task Force Headquarters". His reasoning was it would have been slower for the Army to do things rather than have Darwin's civilians do them.

Stretton's visit to the Army depot, as cited, produced another complaint from him: a surplus officer:

During my visit, I also discovered that Larrakeyah Barracks housed not just one Lieutenant-Colonel, but two! The Officer Commanding the Army was due to be relieved on 1st January and his relief had arrived on 16th December. The two of them had been there before, during and after the cyclone. How I could have used this spare officer during those initial hectic days! I do not intend to criticise the behaviour of those who went through so much. I was fortunate in not experiencing the trauma of Tracy – my sincere sympathy rests with those who did.

These complaints did not make their way into the public domain at the time, or even immediately afterwards. But it was quick enough to see the light of day. Stretton's book *The Furious Days* was published in 1976. He did not confine himself to criticising just the Army: the lack of both Territory and federal foresight and preparation in his eyes were also given liberal negative comments. But they did not go unanswered. It is outside the scope of this work to analyse them all, but it is well known, and "return fire" to these can be found readily enough in the newspapers and journals of the time. But Stretton's comments in his book about the Army were met with some furious and detailed response.

Amongst them were Denis Castle's letter in the *Canberra Times*, wishing to refute an article which had appeared in *The Bulletin*, a national weekly magazine:

LETTERS to the Editor

Sir, — I write in relation to Major-General Stretton's "expose" of Darwin "blunders" as published in *The Bulletin* of November 6. I ask you to publish my comments in the interests of fair play and decency.

At the time of Cyclone Tracy I was Chief Transport and Movements Officer, commanding Seventh Transport and Movements Group in Darwin. My responsibilities included the control and co-ordination of all Army movement, including removals and Army transport in the Northern Territory.

Major-General Stretton has maligned me, my unit, and the Army as it was in Darwin at the time. His comments reflect on the integrity and professionalism of the whole Australian Army. I accuse, Major-General Stretton of allowing to be published, using his name, inferences, half-truths and false accusations as to the conduct of the Australian Army in Darwin at the time immediately following Cyclone Tracy.

I refute the article in *The Bulletin* on the grounds of personal knowledge of a number of aspects both mentioned and not mentioned. I accept one point, that we neither anticipated nor prepared for a major disaster which resulted from winds of 200 mph plus. Would, could, anyone?

On communications, where was mention of the fact that the first outside link was established, after all the aerials had blown away, from the back of an Army Landrover to a signals unit in Townsville? Did not Major-General Stretton's National Disasters Organisation receive information as to the situation via this link?

The inference is made that the Army and the RAAF were looking after their own interests in evacuating their own dependants. It must be understood that this decision was taken by service commanders at least 36 hours before the general evacuation was announced.

In the case of the Army, a number of factors contributed, including the fact that not one married quarter in Larrakeyah Barracks was habitable, and dependants were being accommodated in the three mess buildings.

On Boxing Day, RAAF aircraft were already arriving with relief supplies, and would have been returning empty if space had not been utilised for [dependants of] Soldiers of the Seventh Transport and Movement Group unloading Bailey bridging from Navy landing craft at Darwin to repair damage caused to the wharf by Cyclone Tracy.

The Evacuation Committee had not yet been activated as the first service dependants left, at about 0930 on Boxing Day. By 0400 on 27 December, the last major airlift of Army dependants had left, allowing all soldiers to be fully committed to the aid of the rest of Darwin.

Many wives did not want to leave their husbands and their scattered possessions, some were without relatives or friends to go to, but all were ordered to leave. There was no major in charge of supplies and transport. There was a major in supplies, and I was in charge of transport. Sixteen 2½-tonne vehicles had not been engaged on camp duties on

December 25, 26 or 27. Only two of these vehicles were mobile, and the rest were being processed by workshops personnel, working without a break, in order to effect immediate post-cyclone repairs.

It was in response to a request from the Darwin Committee of the Natural Disasters Organisation that all available vehicles were being repaired, and pooled under my command, for relief operations. Two were committed on December 27 to carry soldiers to deserted civilian houses to clear rotting food from refrigerators, freezers and surrounds.

No relieving convoy arrived on the doorstep on the Food Distribution Committee. I did, and there began, after discussions as to priorities, a program of food-warehouse clearance, distribution to refugee centres, wharf and airfield clearance.

As a de facto member of the committee, I worked side-by-side with other members and various co-opted assistants with a harmonious and highly successful relationship which continued for a considerable period of time.

Major-General Stretton next remarks disparagingly on his observations of Larrakeyah Barracks on December 30. It is true that some soldiers had been stood down for a half day at the time of his visit. My unit, for example, had been involved since Christmas Day on the evacuation of dependants, the repairing of vehicles, the carriage of other soldiers to the food-clearing task, the clearance of the airfield, the clearance of the wharf, the food-distribution task and even the placing of a tarpaulin over the vehicle shelter to waterproof a unit sleeping area.

My unit had averaged something like two or three hours sleep each 24 hours for the past 96 hours. We were forced to rest them, and allow a few hours for all of us to clear our own refrigerators and freezers. We spent those few hours dry retching whilst scraping up our maggot-infested Christmas hams, frozen chickens, etc. After five days exposed to the Darwin wet season, they weren't very palatable.

I am sure that Major-General Stretton worked hard for his six days in Darwin. He left on December 31. Some of my soldiers had already left to rejoin their families by the time I left on March 6. Some had not yet been relieved.

I don't know or understand why Major-General Stretton chose to denigrate his service. I feel bound to deny his allegations on behalf of myself, my men, my corps and the Army. They are unjust allegations.

We, too, were residents of Darwin. We, too, suffered the horror of the cyclone. We, too, lost the possessions of a lifetime. I claim no more and no less for my soldiers that I do for all the other people of Darwin: in adversity, they were magnificent; they worked their guts out.

(Captain) DENIS CASTLE, Puckapunyal, Victoria.

Another angry writer of a letter to a newspaper wrote:

If Alan Stretton had the "steely eyes" of an Army major-general and all the authority of the Australian Government, and was supreme commander, it seems strange that he was unable to give leadership to a few exhausted soldiers.

Another clash was over Stretton taking soldiers away from duties they were already assigned to:

On 28 December, DGNDO directed that all rotting food in refrigerators was to be collected and destroyed to prevent an outbreak of disease. LTCOL Rogers, who either attended or was represented at all DGNDO conferences from 7pm CST on 26 December, later stated:

> At the evening conference he (DGNDO) requested that Army be responsible and I agreed. I had no prior warning of this, although I found later that the matter had been discussed without my knowledge prior to the conference. In accepting the responsibility I made the point strongly that the task would require more labour than I had available at Larrakeyah. I again tried to raise with him the matter of LTCOL Burnett's duties, but he would not discuss the matter. Operation "Clean Up" began on 29 December, using Service and civilian labour and Public Health Inspectors.

> This task could only be carried out by withdrawing men from other tasks, such as vehicle repair, maintenance and driving.

Stretton was indeed a polarising figure. Ask those who were present in the Tracy aftermath and they will have an opinion of him. But in the end his legacy is that Darwin was rescued, and it is most unlikely that could have been done without him.

The incredible force of Cyclone Tracy has bent over this metal power pole.

LIST OF WORKS CONSULTED

The author began research on Cyclone Tracy in his own early days in the Navy, when he was tasked, given his civilian writings of military history books, with compiling a history of the RAN's efforts. Hence some sources are listed as being from those days, reflecting contact made with personnel who were present during and after the cyclone. The author also published *Wrecks in Darwin Waters* (Turton & Armstrong) in 1990, which was the first comprehensive examination of shipwrecks caused by Cyclone Tracy.

The original manuscript is heavily footnoted as to the origin of the information. Readers who wish to ascertain such evidence are welcome to contact the author through the publisher.

ABC News. "Dr Ella Stack, mayor who guided Darwin through Cyclone Tracy aftermath, dies aged 94." 22 May 2023.

ABC–TV. Film. *When Cyclone Tracy came to Darwin | That Christmas*. 2021.

Adastron aviation history website, Aviation in the Aftermath of Cyclone Tracy: https://www.adastron.com/cyclone-tracy/default.htm

Anderson, Jan. Evacuee. Interview 29 Feb 2024.

Archibald, Jared OAM. History Curator, Museum and Art Gallery of the Northern Territory. Emails and conversations 2021-2024.

Argirides, Lieutenant Andrea, RANR. "Women in the RAN: The Road to Command at Sea".

Australia – a Baz Lurhmann film. Movie released November 2008.

Australian Army. Newspaper. "A stormy way to breeze through an Aussie tour." 20 Mar 1975.

Australian Army. Newspaper. "The continuing story of the relief and reconstruction of Darwin." 3 Apr 1975.

Australian Broadcasting Commission. "Cyclone Tracy recovery leader Stretton dies." 28 Oct 2012.

Australian Broadcasting Commission. Film. *The Darwin Story*. 1975.

Australian Broadcasting Corporation. AM (Radio program) "NT coroner hands down finding on Cyclone Tracy deaths." Original broadcast 18 March, 2005.

Australian Defence Force. Film. *Operation Navy Help*. Defence Public Relations.

Australian Defence Force. "Pay & Entitlements." Accessed June 2022.

Australian Dictionary of Biography. Damien Murphy. "Alan Bishop Stretton (1922–2012)."

Australian Federal Police. *On Duty*. Canberra: 2013. Accessed September 2021.

Australian Government. Department of Veterans Affairs. Muir, William Daniel. WWII Post-nominal roll.

Australian Government. Attorney-General's Department Disasters Database.

Australian Government. Attorney-General's Department Disasters Database. Australian Emergency Management Institute. 2012.

Australian Government. Department of Science. Bureau of Meteorology. *Report on Cyclone Tracy 1974*. Australian Government Publishing Service, 1977.

Australian Government. Gissing, Andrew. Australian Disaster Resilience Knowledge Hub. "Leading through crisis: the leadership experience of Major General Alan Stretton." Australian Journal of Emergency Management. April 2022 edition." Accessed Dec 2022.

Australian Government. Department of the Prime Minister and Cabinet. (Reference to Bob Dagworthy)

Australian Parliament. House of Representatives. 29th Parliament, 1st Session. 11 February 1975.

Australian Parliament. Federation Chamber Private Members' Business – Cyclone Tracy. Speech on 40th anniversary. 1 December 2014.

Australian Parliament. House of Representatives. Table Office, Department of the House of Representatives. Emails February 2024.

Australian Naval Aviation Museum. *Flying stations: a story of Australian naval aviation*. NSW, 1998.

Blanch, Paul, Lieutenant Commander, conversation with the author. 19 August 1994.

Bloomfield, Peter, Chief Petty officer, letter to the author. October 1994.

Brown, Malcolm. *Australia's worst disasters*. Victoria: Lothian Books, 2002.

Blue, RS, Lieutenant Commander, RAN. *United and Undaunted*. The History of the Clearance Diving Branch of the RAN. Sydney: Naval Historical Society of Australia, 1976.

Bradford, John. Report commissioned by the Darwin City Council. "Inquiry into enemy air raids on Darwin." War Cabinet Agendum No.116/1942. NAA Series No. A5954, Control Symbol No. 524/4, page 120. 2009.

Bunbury, Bill. *Cyclone Tracy: picking up the pieces*. WA: Fremantle Arts Centre Press, 1994.

Business Council of Australia. "Sir Peter Cosgrove interview on 3AW Mornings with Neil Mitchell." 25 February 2020.

Cannon, Michael. *Australia: A history in photographs*. Victoria: Currey O'Neil, 1983.

Castle, Denis. "Letters to the Editor." *The Canberra Times*. 17 Nov 1976.

Coe, John J. *The Territory: an historical perspective*. Darwin: Museum and Art Galleries Board, 1981.

Commonwealth Bank Manager to Governor of the Bank. Letter. "Evacuation of Darwin Branch." NSW. 24 March 1942. Alice Springs. (Copy in possession of the author, provided courtesy of Jack Mulholland)

Commonwealth of Australia. *The Experience of Cyclone Tracy*. Canberra: Australian Government Publishing Service, 1981.

Corby (nee Lowe), Yvonne, ex-RAN, conversations with the author. July-November, 1994.

Cosgrove, General Sir Peter. *You shouldn't have joined...A memoir*. NSW: Allen and Unwin, 2020.

Cunningham, Sophie. *Warning: the story of Cyclone Tracy*. Victoria: Text Publishing, 2014.

Cunningham, Sophie "Descended upon by looters." *Overland Magazine*. Issue 209, Summer 2012.

Cyclone Tracy. Remembering Cyclone Tracy, fifty years later (Website) "Remembering those who died."

Dadswell, Commodore Thomas, AM, RAN (Rtd), emails and interview with the author, 2023.

Dagworthy, Captain Robert, AM RAN (Rtd). Report written for Sea Power Centre, 2021.

Dagworthy, Captain Robert, AM RAN (Rtd). Emails and conversations with the author, 2022-2023.

Darwin City Council plaque of Cyclone Tracy fatalities. Council Chambers Darwin.

Densten, Frank, Lieutenant Commander, RAN, letters to the author, and reports, August, 1994.

Department of Defence. *The Defence Force in the Relief of Darwin after Cyclone Tracy*. Australian Government Printing Service, 1980.

Department of Housing and Construction. George R Walker. "Report on Cyclone "Tracy Effect on Buildings." Department of Engineering. James Cook University of North Queensland, 1975.

Digney, Larry. *Bubbles, Booze, Bombs and Bastards, A Clearance Diver's Story*. Hobart: self-published, 2023.

Dimond, Glenys. *Cyclone Tracy: an unforgettable Christmas*. Darwin: Museum and Art Gallery of the NT, 2004.

Dixon, Dr Lorraine. Interview, March 2024.

Director-General Natural Disasters Organisation. *Darwin Disaster: Cyclone Tracy*. Canberra: Australian Government Publishing Service, 1975.

Discussion in the "Old Darwin" Facebook group between the author and son Michael Pastrikos, September 2023.

Dos Santos, Paula. NT Archives. NTRS 1148 Personal papers relating to Darwin, the Navy and Cyclone Tracy 1962-1975.

Dowson (*nee* Crosswell), Virginia, Warrant Officer, 21 July 1994 and other dates, conversations with the author.

Eames, Jim. *Taking to the skies: daredevils, heroes and hijackings Australian flying stories from the Catalina to the Jumbo*. Sydney: Allen & Unwin, 2014.

Edwards, Peter. *Arthur Tange: the last of the Mandarins*. Crows Nest, NSW: Allen & Unwin, 2006.

Facebook group. Cyclone Tracy Survivors. Comments. 11 Mar 2022.

Fanderlinden, Flying Officer Jack. Email to the author, Feb 2024.

Film Australia. Film. *Cyclone Tracy – Darwin, Christmas, 1974*. Directed by Chris Noonan. 1975

Film Australia. Film. *When Will The Birds Return?* 1975.

Foster, Bob, and Renegade Renehan. *Beyond Birdum. South Australia: Seaview Press, 2008*.

Frame, Ian. Former RAAF Officer. Emails to the author, February 2024.

Gibson, Eve. *Beyond the boundary: Fannie Bay 1869-2001*. Darwin: Historical Society of the NT, 2011.

Government House Northern Territory. Website.

Griffiths, Rear Admiral Guy, AO, DSC, DSO, RAN (Rtd.) Interview, 2 June 2022.

Haldane, Robert. *The People's Force: a history of the Victoria Police*. Carlton: Melbourne University Press, 2018.

Hancock, Dr. Jonathon Yeatman. NT Archives. NTRS 226 Oral history interview transcript 848.

Handmer, John, and Pascale Maynard. "Civil society mobilisation after Cyclone Tracy, Darwin 1974." 2020. Environmental Hazards, DOI: 10.1080/17477891.2020.1838254

Hitchins, Air Commodore David. NT Archives. NTRS 226.

Johnston, Eric Eugene. "Operation Navy Help: Disaster Operations by the Royal Australian Navy, post Cyclone Tracy". Northern Territory Library Service, Darwin. 07 Jul. 1986 Web. 24 Feb. 2022.

Johnston, Eric, Commodore, RAN. (Rtd.) NOCNA during Tracy, Interview with the author. 2 August 1994.

Jones, Peter D. *Guy Griffiths: the life & times of an Australian Admiral*. North Melbourne, Vic: Arcadia, 2021.

Jones, Dave, Lieutenant Commander RAN. Interview 2 Feb 2022.

Jones, Dave. Sergeant loadmaster with 36 Squadron RAAF. Interview, February 2024.

Katherine Region of Writers. *A slice of Territory life*. Katherine, 2013.

Kendall, Stephanie. (Ed.) *Remembering Cyclone Tracy 24 December 1974. Darwin:* Council on the Ageing (Northern Territory). 2014.

Keogh Jayne. "HMAS *Arrow* in the eye of the storm." Naval Association of Australia. 9 December 2021.

Kershaw, Keith. *Aviation in the Aftermath of Cyclone Tracy*. "Recollections Of Keith Kershaw C-130 Hercules Flight Engineer 37 SQN RAAF." Website. January 2012.

Lamb, John. *Silent pearls: old Japanese graves in Darwin and the history of pearling*. Self-published, 2015.

Lawrie, Dawn. NT Archives. NTRS 226 Oral history interview transcript 505.

Ledger, Geoff, Commander RAN, letter to the author, August, 1994.

Lewis, Tom. *Wrecks in Darwin Waters*. Sydney: Turton and Armstrong, 1990.

Lewis, Tom. *The Empire Strikes South*. Adelaide: Avonmore Books, 2017.

Lewis, Tom. *Bombers North* – a history of bombers operating out of Australia in WWII. Avonmore Books, 2023.

Lewis, Tom. *Eagles over Darwin*: the USAAF defending northern Australia in 1942.Avonmore Books, 2021.

Lewis, Tom. *Teddy Sheean VC*. Sydney: Big Sky, 2021.

Lewis, Tom. *Darwin's Submarine I-124*. Avonmore Books, 2010.

Lewis, Tom. *By Derwent Divided*. Darwin: Tall Stories, 1999.

Lewis, Tom. "What Fairmile is that?". Article published in numerous journals.

Library and Archives NT. "Explore_NT_History_Cyclone_Tracy_20180522". 22 May 2018.

Lowe, Yvonne, senior WRAN, (later Yvonne Corby). Interview. July 1995.

Luxton, Bob, RAN (Rtd.) Interview, 14 March 2022.

Mallyon, David, Warrant Officer, 21 July 1994, conversations with the author. 1994.

Manzie, Daryl, AM. Former NT police sergeant. Interview 19 June 2023, and various emails and conversations.

Markwell, Ken. Interview with the author. February 2022.

Marshall, Ian. NT Archives. NTRS 226 Oral history interview transcript 489.

McHenry, Ray. NT Archives. NTRS 226 Oral history interview transcript 270.

McKay, Gary. *Tracy: the storm that wiped out Darwin on Christmas day 1974*. NSW: Allen and Unwin, 2004.

McLaren, Bill. Library & Archives NT, Northern Territory Archives Service, NTRS 226, Typed transcripts of oral history interviews with 'TS' prefix, TS 586.

Mitchell, Brett. Royal Australian Navy. "Disaster Relief – Cyclone Tracy and Tasman Bridge".

Monument Australia. "Cyclone Tracy Memorial".

Muir, Hilda Jarman. *Very big journey: my life as I remember it*. Canberra: Aboriginal Studies Press, 2004.

Murphy, Kevin. *Big Blow up North* (A History of Tropical cyclones in Australia's Northern Territory). Darwin: NT Government (University Planning Authority), 1984.

Museum and Art Gallery of the NT. TH99/046. Department of Social Security'. Booklet – 'The Experience of Cyclone Tracy.

Museum and Art Gallery of the NT. TH99/047.1-.2.1 Booklet – 'The Defence Force in the Relief of Darwin after Cyclone Tracy', (1980) annotated by General Stretton, 300 x 220mm (H x W). .2 Yellow envelope addressed to Major General A. Stretton from the Australian Red Cross Society, 325 x 230mm (H x W).

Museum and Art Gallery of the NT. TH99/054. Booklet – 'The Bulletin, November 6, 1976' (Article – 'The Darwin Blunders Exposed – General Stretton's Own Story')

Museum and Art Gallery of the NT. TH99/061. 4 foolscap pages labelled "Notes from NEOC log, Xmas morning 25 December 1974."

Museum and Art Gallery of the NT. TH99/066.1 (A-N) Photocopy – Typed log from National Emergency Organisation Centre (NEOC), Canberra, 25 December - 30 December 1974 including entry 21 December.

Museum and Art Gallery of the NT. TH99/066.5 (A-B)   A. Photocopy – Hansard House of Representatives 18 November 1976, Mr Killen responding to allegations made by General Stretton in the publication "Furious Days". B. (Continuation of A.) Photocopy – Hansard House of Representatives 18 November 1976, Mr Killen responding to allegations made by General Stretton in the publication "Furious Days".

Museum and Art Gallery of the NT. TH99/067.1. Typed Transcripts – Radio broadcasts made by General Stretton on the emergency frequency, 26 December - 29 December (23 pages stapled together in the top left hand corner).

Museum and Art Gallery of the NT. TH99/067.3. Typed copy – Rough notes, "The General's Last Battle" (Three pages, by Alan Stretton, stapled together in the top left hand corner).

Museum and Art Gallery of the NT. TH99/067.4-.5.4 Copy of Typed Statement by Major-General Alan Stretton, 17 May 1978.

National Australian Archives. "Cyclone Tracy – Navy Help." Series number E499. Item ID 7041035.

National Archives of Australia. "Updated list of deceased persons who came to their deaths during cyclone [Tracy] at Darwin on 25/12/74 identified / not identified." Item number 7068518.

National Archives of Australia. Fact Sheet 176. "Cyclone Tracy, Darwin".

National Emergency Organisation Centre. TH99/066.1 (A-N). Typed log, Canberra, 25 December-30 December 1974 including entry 21 December. (In MAGNT Collection)

National Climate Change Adaptation Research Facility. Haynes, K, Bird, DK, Carson, D, Larkin, S & Mason. Institutional response and Indigenous experiences of Cyclone Tracy. Gold Coast, 2011.

Naval Historical Society of Australia. "Obituary: Commodore EE Johnston AO AM OBE."

Naval Historical Society of Australia. Author given as "A.N. Other". Occasional Paper 169: HMAS Vendetta and Commander Eric Eugene Johnson RAN; Vietnam Deployment 1969-1970. Sep 1, 2023.

Naylor, Jim. RAAF Sergeant. Emails to the author, February 2024.

Nixon, "Curly". NT Archives. NTRS 654. Transcript of oral interview.

Northern Territory Dictionary of Biography. Carment, David. "Nelson, John Norman (Jock)."

Northern Territory Government. "Maritime Heritage".

Northern Territory Government. Inquest into the deaths of Raymond John Curtain, Terrence Carew Westwood, Gerald Frederick Thompson, Gregory John Westerman, Graham William Dearden & Ruth Nazmeena Adrione Vincent. [2005] NTMC 016.

Northern Territory Library. "History of Self-Government." https://territoryday.nt.gov.au/history-of-self-government/

NT News. Peden, Ludij. "Cyclone Tracy: Survivor's story." 11 Jan 2014. Accessed June 2021.

NT News. "Historian denies Bombing toll was higher." 17 February 2012.

NT News. "Donn's record Cyclone Tracy mercy flight." 20 December 2014.

O'Grady, Tom. NT Archives. NTRS 226 Oral history interview transcript 494.

Odgers, GJ. "The Defence Force in the Relief of Darwin after Cyclone Tracy." Australian Government Publishing Service: Canberra, 1980.

Oppenheimer, Melanie. *The Power of Humanity: 100 years of Australian Red Cross 1914-2014.* Australia: HarperCollins Australia, 2014.

Parliament of New South Wales. *Report of the Police Department for 1974.* 1975.

Parliament of the Commonwealth of Australia. The Commonwealth Police Force. *Annual Report of the Commissioner of Police.* Parliamentary Paper No. 237/1976. 30 June 1975.

Pedersen, Lieutenant PA. "A Platoon Commander's Experience." *Army Journal.* No. 316. September 1975.

Post-Courier. "Storm Warning." Papua New Guinea. 4 Dec 1974.

Pratt, Brenda, ex-RAN, conversations with the author, July-November, 1994.

Pratt, Keith, ex-RAN, conversations with the author, July-November, 1994.

Queensland Government. "From the Vault – 40 years since Cyclone Tracy." 23 Dec 2014.

Rayner, Robert. *The Army and the Defence of Darwin Fortress.* NSW: Rudder Press, 1995.

Reid, Bill. Officer, Australian Army. Reply to the author's article in *Quadrant* magazine, and subsequent emails, July 2023.

Riddle, John. NT Archives. NTRS 226 Oral history interview transcript 811.

Risk Management Solutions. "Cyclone Tracy 30-Year Retrospective". 2005.

Roberts, Patti. *Surviving Tracy: Cyclone Tracy survivor stories.* Paradox Promotions, 2015.

Robertson, H. "Darwin's Churchill: the role of Major-General Alan Stretton in the days following Cyclone Tracy." Journal of Northern Territory History, 1999.

Rosenzweig, Paul. "Role of RAN and Captain Eric Johnston in the recovery of Darwin in the aftermaths of Cyclone Tracy which struck on Christmas Day 1974." *Sabretache*: the Journal of the Military Collectors Society of Australia. 1986.

Royal Australian Air Force. Medical Report. (Courtesy Ken Stone). 7 June 1988.

Royal Australian Navy. Mitchell, Brett. "Disaster Relief – Cyclone Tracy and Tasman Bridge". Undated.

Royal Australian Navy Communications Forum. Unnamed ex-RAN member. "My Cyclone Tracy Experience."

Royal Australian Navy. Report of the Board of Enquiry into the Circumstances Attending the loss of or Damage to HMA Ships *Arrow, Attack, Assail* and *Advance* and the casualties resulting therefrom. (Copy courtesy Sea Power Centre, Canberra), 1975.

Royal Australian Navy. *Navy News.* Volume 18, No. 1. 17 January, 1975.

Royal Australian Navy. HMAS *Watson* History.

Royal Australian Navy. "HMAS *Arrow*."

Royal Australian Navy. Port Darwin Harbour Survey 1975. Copy retained by the author.

Royal Australian Navy. Semaphore. *Semaphore: Disaster Relief – Cyclone Tracy and Tasman Bridge.* Issue 14, 2004.

Sayers, Susan. *The Not So "Silent Night": Cyclone Tracy stories from doctors, nurses and health workers.* Darwin: Historical Society of the NT, 2015.

SBS News. "Darwin marks the day Tracy blew Christmas away." 19 December 2014.

Schulz, Dennis. *Tracy Tales: how the Darwin business community survived the great cyclone.* Darwin: Northern Territory Government Department of Business, 2014.

Shackleton, David, Rear-Admiral, Chief of Navy. Anecdote on Tracy related to New Entry Officer Course members, RAN College, 27 February 2001. (Author present)

Simmons, Lieutenant Commander John, RAN, letter to the author, August, 1994. St Leonards, N.S.W: Allen & Unwin, 1998.

Sky News. Film. *Remembering Tracy – 40 Years On.* 2014.

Slater, Dee. *Tracy's Fury.* Self-published: Warwick, Qld; 2018.

Smith, Camden. *NT News.* "New Australian Defence Force boss David Johnston's Northern Territory connection." 10 April 2024.

Stack, Dr Ella. *Is there anyone alive in there: our Cyclone Tracy, Darwin, Christmas 1974.* Darwin: Historical Society of the NT, 2013.

State Library of South Australia. "Cyclone Tracy Evacuees." Undated.

Stone, Ken OAM, and Margaret Stone. Interview with the author, October 2023.

Stretton, Alan. *Soldier in a storm: an autobiography.* Sydney: Collins, 1978.

Stretton, Alan. *The Furious Days.* Sydney: Collins, 1976.

Swan, Captain Brian AM RAN (Rtd.) Letters urging awarding of a medal for the relief operations. Courtesy Admiral Guy Griffiths.

Sydney Morning Herald. "Tracy's final victim." 3 March 1990. Accessed June 2021.

Sydney Morning Herald. "Tracy's final victim." 3 March 1990. Accessed June 2021.

Tambling, The Hon Grant. "Rebuilding Darwin Post Cyclone Tracy". Address to the National Young Planners Conference, Darwin on 28th March 2009. (Supplied to the author)

Tarrier, Stuart. RAAF sergeant. Interview February 2024.

Taulelei, Michael. Evacuee. Interview with the author, February 2024.

The Advertiser. "Cyclone Tracy 40 Years On – Part 1: Impact & Survival – THE HARDEST JOB OF ALL". 25 Nov 2014.

The Canberra Times. "Inaccuracies in book." 17 Nov 1976. (p. 2)

The Canberra Times. "General Stretton raises the question of the public's right to know. Rumbles of another storm." 17 May 1978. (p. 19)

The Canberra Times. "Police deny death toll higher". 11 Jan 1975. (p. 7)

The Canberra Times. "Relief for Darwin Force." 27 Feb 1975. (p. 11)

The Canberra Times. "ACT men search for bodies." 31 Dec 1974. (p. 1)

The Canberra Times. "Seats for Darwin concert rushed." 31 Dec 1974. (p. 1)

The Governor-General of the Commonwealth of Australia. "National Emergency Medal".

The Northern Territory Police Museum and Historical Society. Website: www.ntpmhs.com.au

The Salvation Army. Website. "Hope springs from devastation." Undated story.

This Adventurous Age. Website. Accessed Jan 2022.

Tucker, Alan. *Cyclone Tracy*. NSW: Scholastic Press, 2004.

Victoria Police. Email from Anna Burnett, Historical Research Officer, Victoria Police Museum. Media, Communications and Engagement Department. 23 August 2022.

Whitaker, Richard. *Australia's natural disasters*. Sydney: Reed New Holland, 2006.

Williams, Jacquie. *Cyclone child: Cyclone Tracy in the maternity ward*. Self-published, 2004.

Willowby, Quentin. *Melbourne so far away*. Morpeth, NSW: Christmas Creek Publications, 1981.

Wilson, Beverley. NT Government worker in 1974-5 for the organization of the permit system. Interview September 2023, and emails 2024.

Women's Day. "Cyclone Tracy 30 Years On." 27 December 2004.

Yeoward, Kym. Australian Army. Interview 28 March 2022.

INDEX OF NAMES

Allen, Lieutenant Colonel Bob 72

Anderson, Jan 44

Andrew, Peter 109

Argirides, Andrea 63

Beal, Dr 87

Bell, Malini Palathil 109

Birtchnell, Sub-Lieutenant Andrew 23

Blanch, Lieutenant Commander Paul 24

Bloomfield, Dorothy 109

Bloomfield, Peter 62

Bolger, Wotsanbuk 109

Bonner, Cecil Henry 109

Bowning, Sergeant Graham 97

Brennan, Harry "Tiger" 58

Brown, Christine 109

Brown, Geraldine Elizabeth 109

Brown, Kay 109

Brown, Stephanie 109

Bruhn, Andrew Mark 109

Bullock 99

Burgess, Dean William 109

Burnett, Lieutenant Colonel 120

Butler, Louisa Fanny 109

Cadd, Brian 102

Cairns, Dr Jim 32

Carolan, Sergeant Alex 100, 101

Castle, Captain Denis 118, 119

Catton, Petty Officer Leslie 23, 109

Chaney, Paul Mark 109

Clarke, Rose Susanna 109

Cleveland, Lieutenant Chris 24

Clough, Colin 109

Clough, Eileen Patricia May 109, 112

Conje, Barbara 52

Cosgrove, General Sir Peter 7, 68, 69

Cousins, Eileen 84

Cousins, Inspector Len 84

Cowan, Sergeant 45

Crocker, Barry 102

Cunningham, Sophie 32, 97, 99, 101, 102, 108

Curtain, Raymond John 25, 109

Dadswell, Commodore Thomas 50, 56, 57, 61, 64, 67

Daffey, Peter Brian 109

Dagworthy, Lieutenant Robert 22, 23, 25, 47, 60, 61

Daniel, Avis Winifred 109

Day, Bill 104

Day, Polly 104

de Graaf, Lieutenant Paul 24

Dearden, Graham William 109

Dennis, CJ 71

Densten, Lieutenant Frank 2

Dewar, Peter James 110

Dibua, Charles Joseph 110

Digney, Larry 55

Dixon, Greg 19, 20

Dixon, Lorraine 19, 20

Elizabeth II, Queen 102, 106, 107, 113

Fanderlinden, Flying Officer Jack 38, 39

Farnham, Johnny 102

Fealy, David Grant 110

Fenton, Gary Roger 110

Fewster, Squadron Leader Bill 39

Frame, Ian 41, 88

Goddard, SG 87

Grannall, Father 18

Grant, Michael John 110

Griffiths, Commodore Guy 49, 53, 61

Grose, Petty Officer 61

Gurd, Charles 45

Hammett, Tony 69

Hammond, Dame Joan 102

Hampton, "Happy" Raymond 19, 110

Hancock, Dr Jonathon 59

Hanson, Thomas William 110

Harder, Lisa Karoline 110

Harris, Rolf 101, 102

Hewett, Colleen 102

Heys, Commander G 57

Hitchins, Group Captain D 18, 36, 37, 39, 42, 44, 45

Hoff, Donald James Duncan 110

Hogan, Paul 102

Holten, Dennis Keith 110

Howe, Don 44

Huddy, Barbara 101

Humphries, Barry 102

Jackson, Reg 85

Jacobi, Sub-Lieutenant John 23

James, Richard Roland 110

James, Rueben Paul 110

Johnson, Johno 57

Johnston, Vice Admiral David 20

Johnston, Captain Eric 6, 20, 24, 25, 51-55, 57, 60, 64

Jones, Dave 43, 55

Kalio, Staff Sergeant Eric 72

Kennon, Stan 56

Kentish, Reverend Len 14

Kerr, Sir John 91

Kershaw, Keith 37

Kingsley, Flying Officer Randall 41

Kitchener, Lord 14

Knox, Catherine Michelle 110, 112

Knox, Elva Dawn 110, 112

Knox, Michael John   110
Knox, Stephen James   110, 112
Lai, On'ing Owen   110
Lang, John Dudley   110
Latter, Bob   97
Lawrie, Dawn   59
Ledger, Geoff   54
Lim, Arthur Fong   108, 110, 113
Lim, Katrina Fong   110
Lowe, Yvonne   20, 26
Luxton, Bob   5
Macklin, Paul Alaister   111
Malley, Fiona   99
Malley, Joyce   99
Malley, Sergeant Kevin   98, 99
Malley, Stephen   99
Manzie, Daryl   84, 101, 105
Markwell, Ken   25, 44
Marshall, Ronald Charles   110
McHenry, Ray   92
McKay, Gary   102, 107
McLaren, Commissioner Bill   83, 85, 87, 92, 101-103, 106
McLeod, Able Seaman Robert   23
McNab, Eric Arthur   111
Mitchell, Brett   106
Monaghan, Wing Commander WJ   18, 35
Muir, William Daniel   111
Mulholland, Jack   15
Murphy, Helen   87
Murphy, Kevin   106
Murphy, Senator Lionel   91
Myers, John   116
Naylor, Jim   40
Nelson, John "Jock"   32, 33
Newton-John, Olivia   102
Nunn, Judy   103
Odawara, Shigemori   111, 112
Parker, George Valentine   111
Pastrikos, Steve   26
Patterson, Dr Rex   36
Peden, Ludij   40, 45
Pedersen, Lieutenant Peter   73
Perry, Roland   103
Portman, Suzanne Mary   111
Pye, Dudley   81
Purcell, Corporal   18
Rainbow, Kevin   23
Reed, Genevieve   113
Reid, Bill   80
Rennie, Able Seaman Ian   23, 111
Riddle, John   58
Robbins, Corporal David   73
Robey, Air Vice Marshal   37, 39

Roewer, George   105, 111
Rogers, Lieutenant Colonel RB   66, 120
Rosenzweig, Paul   52
Sandford, Flying Officer Carl   39
Sheean, Teddy   14
Simmons, Lieutenant Commander John   48, 57
Slater, Dee   98, 100, 101, 103
Smaniotto, Giovanni   111, 112
Smith, Major Donald Macleod Yeomans   81
Spencer, Petty Officer   61
Stack, Dr Ella   91
Stephenson, Cherry Leona Rose   100, 111, 112
Stephenson, Able Seaman Geoffrey   25, 99, 111
Stephenson, Kylie Jane   100, 111
Stone, Flight Sergeant Ken   18, 19
Stone, Margaret   18, 19
Stretton, Major General Alan   30-33, 36, 42, 51, 52, 67, 81, 83, 87, 90-93, 100, 101, 105, 114-120
Strickland, Captain   18
Swann, Robert Gordon   111
Tambling, Grant   26
Tarrier, Stuart   39, 40
Taulelei, Angie   44
Taulelei, George   44
Taulelei, Mick   44
Taulelei, Phillip   44
Taylor, John Thomas   111, 112
Thompson, Gerald Frederick   111
Thompson, Richard   111
Thorogood, Major Frank   32, 92, 117
Turner, John   111, 112
Viani, Captain Mike   70, 71
Vincent, Ruth Nazmeena Adrione   111
Wade, Robert Norman   111
Wauchope, Sergeant Trevor   105
Wells, Rear Admiral DC   48, 64
Wenck, Shirley   111
Westerman, Gregory John   112
Westwood, Terrence Carew   112
Wheatley, Kenneth James Scott   100, 112
Whitlam, Prime Minister Gough   30, 91, 93
Williams, Kerry Lynda   112
Williams, Molly   112
Wilson, Beverley   102
Wilson, Ron   88
Wolfe, Captain Kevan   72
Wolfe, Siegfried Karl Otto   112
Wood, Jennifer Anne   112
Woodyatt, William   112
Yeates, Senator Herbert   103
Yeoward, Kym   59
Yoshida, Koji   112
Yu, Frederick Yee Char   112